Digitized Labor

Lorenzo Pupillo · Eli Noam
Leonard Waverman
Editors

Digitized Labor

The Impact of the Internet
on Employment

Editors
Lorenzo Pupillo
Centre for European Policy Studies
Brussels, Belgium

Eli Noam
Columbia Institute for Tele-Information
Columbia Business School
New York, NY, USA

Leonard Waverman
DeGroote School of Business
McMaster University
Burlington, ON, Canada

ISBN 978-3-319-77046-8   ISBN 978-3-319-78420-5   (eBook)
https://doi.org/10.1007/978-3-319-78420-5

Library of Congress Control Number: 2018937855

Cover design by Tom Howey

Printed on acid-free paper

This Palgrave Macmillan imprint is published by the registered company Springer International Publishing AG part of Springer Nature
The registered company address is: Gewerbestrasse 11, 6330 Cham, Switzerland

To the Next Generation

Acknowledgements

The original impetus for this book came from a conference on the Impact of the Internet on employment held at Columbia Institute for Tele-Information. Since then this topic started to become part of the CITI research. We would like to thank all of those who contributed chapters to this volume. A special thank goes to the International Development Research Centre Canada and the Department for International Development (DFID) UK for supporting the research on "Investigating the potential for Micro-work and Online-freelancing in Sri Lanka" and to Anne Harrington and Margarita Minkova for editorial assistance.

Contents

Editors and Contributors

About the Editors

Dr. Lorenzo Pupillo is currently an Associate Senior Research Fellow at the Centre for European Policy Studies in Bruxelles and Head of the Cybersecurity@CEPS Initiative. His research is focused on: the impact of digital technologies on jobs, the economics of cybersecurity and the dynamics of Internet Governance. He has more than 30 years of experience in the private sector, government, international organizations and academia. Before joining CEPS, he served as an Executive Director in the Public & Regulatory Affairs Unit of Telecom Italia developing the company' global public policies for Internet, Cyber-Security, Next Generation Networks. He also managed Telecom Italia's relations with the OECD, the ITU and other international associations and organizations. Previously, Dr. Pupillo held a variety of senior positions in the Strategy, Business Development and Learning Services divisions of Telecom Italia. He is an economist by training and has worked in many areas of telecommunications demand and regulatory analysis, publishing four books on Internet Policy and many papers in applied econometrics and industrial organization. He has served as an advisor to the Global Information and Communication Technologies Department of the World Bank. Before joining Telecom Italia, he was member of the technical staff at AT&T Bell Laboratories in Murray Hill—New Jersey—and he worked as senior economist for governmental institutions. Dr. Pupillo is also an affiliated researcher at Columbia Institute for Tele Information at Columbia Business School and serves on numerous scientific and advisory boards around the globe. He is also adjunct Professor of

Global Governance of Digital Technologies at the University of Urbino. He obtained a Ph.D. and an MA from University of Pennsylvania, an MBA. from Istituto Adriano Olivetti in Ancona Italy and an MS in Mathematics from University of Rome.

Eli Noam Professor of Economics and Finance at the Columbia Business School since 1976, and its Garrett Professor of Public Policy and Business Responsibility. Served for three years as a Commissioner for Public Services of New York State. Appointed by the White House to the President's IT Advisory Committee. Director of the Columbia Institute for Tele-Information, a research center focusing on management and policy issues in telecommunications, internet, and electronic mass media. He has also taught at Columbia Law School, Princeton University's Economics Department and Woodrow Wilson School, and the University of St. Gallen, and is active in the development of electronic distance education. Noam has published 29 books and over 350 articles in economics journals, law reviews, and interdisciplinary journals, and has been a regular columnist for the *Financial Times* online edition. His recent books include *Broadband Networks and Smart Grids* (Springer, 2013); *Media Ownership and Concentration in America* (Oxford, 2009); *Peer to Peer Video* (Springer, 2008); and *Ultrabroadband* (IDATE, 2008). He is completing a 3-volume series on Media Management, and leads a 30-country team on international media ownership. He is Chairman of the International Media Management Academic Association. Noam has been a member of advisory boards for the Federal government's telecommunications network, and of the IRS computer system, of the National Computer Systems Laboratory, the National Commission on the Status of Women in Computing, the Governor's Task Force on New Media, and of the Intek Corporation. His academic, advisory, and non-profit board and trustee memberships include the Nexus Mundi Foundation (Chairman), the Oxford Internet Institute, Jones International University (the first accredited online university), the Electronic Privacy Information Center, the Minority Media Council, and several committees of the National Research Council. He served on advisory boards for the governments of Ireland and Sweden. Noam is a Fellow of the World Economic Forum, a member of the Council on Foreign Relations, and a commercially rated pilot. He served in the Israel Air Force in the 1967 and 1973 wars, and is an active search and rescue pilot with the US Civil Air Patrol (1st Lt.). He received the degrees of BA, MA, Ph.D. (Economics) and J.D. from Harvard University, and honorary doctorates from the University of Munich (2006) and the University of Marseilles.

Leonard Waverman is currently Dean of the DeGroote School of Business at McMaster University since January 2013. Dr. Waverman is a world-renowned expert in international telecommunications and global resources management. He specializes in microeconomics and industrial organization, economics of telecommunications, energy and resource economics, international trade, public utility, and public enterprise economics. His current research focuses on the impact developments in the telecommunication industry have on growth and productivity. He has authored the influential Connectivity Scorecard, an annual index that ranks countries according to how advanced their communications networks are in promoting productivity and growth. Dr. Waverman has consulted widely on energy, natural resources, telecommunications, and competition policy matters in Canada, the United States, and Europe. In January 2009 he was cited as one of the world's top 50 most influential thought leaders in the telecommunications industry by Global Telecoms magazine and has received the honor of Chevalier dans l'Ordre des Palmes Academiques from the Government of France. He earned his B. Com. and MA from the University of Toronto and his Ph.D. in economics from MIT. He has been a professor of economics at the University of Toronto and the London Business School and Dean of the Haskayne School of Business as well as professor of strategy at the University of Calgary.

Contributors

Dr. Robert D. Atkinson is one of the country's foremost thinkers on innovation economics. With an extensive background in technology policy, he has conducted ground-breaking research projects on technology and innovation, is a valued adviser to state and national policy makers, and a popular speaker on innovation policy nationally and internationally. Before founding ITIF, Atkinson was Vice President of the Progressive Policy Institute and Director of PPI's Technology & New Economy Project. He received his Masters in Urban and Regional Planning from the University of Oregon and was named a distinguished alumnus in 2014. He received his Ph.D. in City and Regional Planning from the University of North Carolina at Chapel Hill in 1989.

Seth G. Benzell is a Postdoctoral Associate at the MIT Initiative on the Digital Economy. He received his Ph.D. in Economics at Boston University, where his adviser was Laurence Kotlikoff. His work focuses on the economics of technological change and its impact on labor markets and welfare.

He is also interested in public finance, the economics of networks, and macroeconomics generally.

Federico Biagi is Associate Professor in Public Economics at University of Padova, Italy. He is currently working as an economist in the Joint Research Center (European Commission) in the Human Capital and Employment Unit. His research interests cover: Technology and the labor market, Human capital and education, R&D and growth. Analysis of income distribution. Economic analysis of the Welfare State. Fiscal federalism, Regulation and Antitrust. He earned a Ph.D. In Economics from University of British Columbia (Vancouver, Canada) and a Ph.D. In Public Finance from the University of Pavia (Pavia, Italy).

Richard Clarke brings both theoretical and practical experience to the study of telecommunications markets. Upon joining Bell Laboratories in 1986, he modeled the likely competitive effects of early proposals to eliminate the Regional Bell Holding Company line-of-business restrictions from the Bell System divestiture consent decree. After moving over to AT&T in 1989, Rich became responsible for AT&T's regulatory policy on access charges, LEC price cap regulation and interconnection rules. From the mid-1990s through 2005, he was responsible for AT&T's economic public policy related to the provision and regulation of competitive local telephone services. This included AT&T's positions on the efficient pricing of interconnection, unbundled network elements, and the costing of universal service. He also directed AT&T's participation in the development of the HAI/Hatfield Model of forward looking economic costs of local exchange networks. Since the acquisition of AT&T by SBC in 2005, Rich has worked on issues related to IP and wireless networks such as network neutrality, video service entry, reverse auctions for universal service, retirement of the PSTN, spectrum auctions and exhaust. In addition, Rich collaborates with the OECD to develop methods to better evaluate the comparative costs and performance of evolving mobile and broadband services in the United States and foreign countries. Rich is the author of numerous papers on economics and telecommunications. He has an A.B. degree in mathematics and economics from the University of Michigan, and A.M. and Ph.D. degrees in Economics from Harvard University. Prior to joining AT&T-Bell Laboratories, he was an Assistant Professor of Economics at the University of Wisconsin-Madison, and worked as an economist at the Antitrust Division of the US Department of Justice.

Robert B. Cohen is a senior fellow at the Economic Strategy Institute where he heads a new study, "The economics and business impacts of the "New IP," on cloud computing, SDN, NFV, big data, and the Internet of

Things. This project will evaluate whether cloud computing and the concomitant reorganization of the workplace will result in revenue gains, productivity improvements and employment and output benefits. In pursuing this analysis, Dr. Cohen is also contributing to the OECD's digital data innovation initiative and the Ewing Marion Kauffman Foundation's work on entrepreneurship and jobs. He is a long-term participant in the Innovation for Jobs organization headed by Vint Cerf and David Nordfors. Dr. Cohen has previously served as Director of TM Forum's Enterprise Cloud Leadership Council. He created New York State's first high tech strategy and has consulted to the European Commission's Directorate General XIII on the internet and economic growth. He is a past president of the Forecasters Club of New York. Dr. Cohen holds an MA and Ph.D. in economics from the New School for Social Research and a BA from Swarthmore College. He is the author, co-author, or co-editor of five books.

Amelia De Rosa is currently head of HR Ecosystem and Partnerships within Telecom Italia's HR Dept—with the responsibility to establish partnerships with universities and research centres for the introduction of new education models, internal know-how sharing and implementing employer branding activities and new initiatives. Amelia has a master degree in Political Science achieved at Luiss Guido Carli University and a Master on advanced service industry. She has an international background, having studied and worked abroad for many years. She has worked in the Regulatory Affairs department based in Rome and Brussels and was responsible for Social and Institutional Projects in Tim Brasil, before heading in the HR department and covering different roles as a responsible of Partnerships and Research, Knowledge Management, Ecosystem and Partnerships, including Employer Branding activities.

Helani Galpaya is CEO of LIRNEasia, a pro-poor, pro-market think tank working across the emerging Asia Pacific on ICT policy and regulatory issues. She researches and engages in public discourse on issues related to net neutrality, policy and regulatory barriers in Internet access, e-Government, broadband quality of service, how knowledge and information disseminated via ICTs can improve inclusiveness SMMEs (small, medium, and micro enterprises) in agriculture and micro-work markets. A primary thrust of her current research is on digital labor in South Asia, where she's just completed a project to understand issues of marginalisation and economic development of those participating in online platforms that enable freelancing and micro-work in India, Sri Lanka, and Myanmar. She is also carrying out nationally representative surveys of Internet use by households and individuals in

India, Pakistan, Bangladesh, Indonesia, Cambodia, Myanmar, and Nepal. She's examining how experiences and perceptions of harassment, surveillance and privacy impact how people of Myanmar experience the Internet, and how tariff structures and tariff changes impact energy poverty and economic development in Sri Lanka. She has been a consultant for the World Bank, UNCTAD, UNESCO, and other organisations on issues related to ICTs and development. Prior to LIRNEasia, she worked on at the ICT Agency of Sri Lanka implementing e-Government projects under the World Bank's e-Sri Lanka initiative. She was a management consultant at Booz&Co.(now Strategy&) in New York and has also worked at Citibank and Merrill Lynch in the USA. She has an MS in Technology & Policy from the Massachusetts Institute of Technology, Cambridge, USA and a BA (Cum Laude) in Computer Science from Mount Holyoke College, Massachusetts, USA.

Giovanni Andrea Iapichino born 1961, has a cultural background on law & economics and during his career has been developing a focus on human resources management in international context and change management. In 2015, he became the Lead member of the Smart working Project in TIM, currently is the company's Welfare Manager pursuing people's inclusion, work-life integration and productivity in the digital transformation era.

Dr. Raul L. Katz is Director of Business Strategy Research at the Columbia Institute for Tele-Information. He is also President of Telecom Advisory Services (www.teleadvs.com), a firm that advises technology clients in the fields of strategy, regulation and business development. During his career, he has worked extensively in the planning and development of digital businesses, particularly telecommunications. In the last five years, Dr. Katz has been focused on analyzing the economic impact of broadband, and the development of national broadband plans. He has led projects on the economic impact of broadband in the United States, Germany, Switzerland, Spain, Senegal, Cote d'Ivoire, Morocco, Panama, Philippines, Colombia, Mexico, Costa Rica, and Ecuador.

Guillermo Lagarda is Senior Economist at the Inter-American Development Bank and Research Fellow at Boston University Global Economic Governance Initiative. Previous to pursing his Ph.D. in Economics at the Boston University he worked for the World Bank for the Latin American and the Caribbean Poverty Reduction Management. His main research interests include the digital economy, public finances, and macro-financial linkages.

Paola Liberace born 1975, is a researcher, a journalist and a communication professional. She gained an extensive experience in media and innovation, working for Mediaset, FullSix and Telecom Italia, where she currently is a Market Researcher and Analyst. As a mother of two, she developed a deep interest in work-life balance and organizational issues, investigating these topics and eventually authoring an essay (Rubbettino, 2009) and several articles. She graduated in aesthetics and philosophy of language at Scuola Normale Superiore in Pisa, and earned a Ph.D. in Communication Sciences at "La Sapienza" University in Rome. A contributor to several newspapers and webzines, she is a professional blogger for Nòva100—Il Sole24ore. Paola is a teacher in regional and national training programs and a lecturer in degree and master courses.

Jonathan Liebenau is Reader in Technology Management at the London School of Economics where he conducts a broad range of research on the digital economy, including studies of internet infrastructure, business practices, and markets. His previous research focused on the pharmaceutical industry, on the history of science-based industry, and on technology and economic development. He holds a doctorate in history and sociology of science from the University of Pennsylvania and is an associate of the Columbia Institute for Tele-Information, Columbia University.

Paolo Naticchioni is Associate Professor at the University of Roma Tre (Department of Political Science and CREI). He received his Ph.D. in Economics from the Université Catholique de Louvain (2008), where he also received a M.Sc. in Economics. He also received a Ph.D. in Economics from the University of Rome La Sapienza (2004). He is the secretary of the Italian Association of Labour Economics (AIEL, since 2016). His research interests include labour economics (in particular topics such as inequality, unions and wage bargaining, wage dynamics, trade and the labour market), urban economics (mainly the analysis of urban wage premia), political economy (political selection and commitment), economics of education, economics of happiness, program evaluation.

Sutharan Perampalam is a senior researcher attached to LIRNEasia. His research interests are mainly on inclusive development, shared economy, financial Inclusion, and decent employment related dimensions. Haran has over 8 years of research experience in managing and conducting quantitative and qualitative researches.

He currently manages research in India and Sri Lanka which explore the opportunities for under-employed youth, women and previously- excluded persons participating in micro-work platforms. Haran was involved in conducting statistical analysis & insight generation for a research that included two nationally representative sample survey of ICT use in Myanmar. He also managed a study looking at policy implications of tariff designing in the electricity sector in Sri Lanka. Prior to joining LIRNEasia, Haran worked as a Research Manager, Consumer Insights at Nielsen and as a Senior Business Analyst at MTI consulting. His business expertise extends to cover different sectors including FMCG, B2B, Industrial and Education for which he carried out many consumer research, customer segmentation, and market entry/feasibility studies.

Giuseppe Ragusa is a senior economist with the European Central Bank. He received his BA and MA in economics from Bocconi University in 1999 and 2000, respectively. He completed his Ph.D. in Economics at the University of California, San Diego in 2005. His first academic job was as an Assistant Professor at Rutgers University from 2005–06. He then taught at University of California, Irvine and at Luiss University in Rome. From 2013 he serves as the director of the Master in Big Data Analytics and scientific director of the Big Data Lab at Luiss Business School. His research interests are theoretical and applied econometrics, labor economics, computational methods, and machine learning applications. The applied part of his research has focused on the study of the labor market with a particular emphasis on patterns of income inequality. He has published many papers on econometrics that have appeared in leading scientific journals including the *Journal of Econometrics*, *Review of Economics and Statistics*, and *Econometric Theory*.

Jeffrey D. Sachs is the Director of The Earth Institute, Quetelet Professor of Sustainable Development, and Professor of Health Policy and Management at Columbia University. He is Special Advisor to United Nations Secretary-General Ban Ki-moon on the Millennium Development Goals, having held the same position under former UN Secretary-General Kofi Annan. He is Director of the UN Sustainable Development Solutions Network. He is co-founder and Chief Strategist of Millennium Promise Alliance, and is director of the Millennium Villages Project. Sachs is also one of the Secretary-General's MDG Advocates, and a Commissioner of the ITU/UNESCO Broadband Commission for Development. He has authored three New York Times bestsellers in the past seven years: *The End of Poverty* (2005), *Common Wealth: Economics for a Crowded Planet* (2008), and *The Price of Civilization* (2011). His most recent book is To Move the World: JFK's Quest for Peace (2013). Professor Sachs is widely considered to be one of the world's leading experts on economic development and the fight against poverty. His work on ending poverty, promoting economic growth, fighting hunger and disease, and promoting sustainable environmental practices, has taken him to more than 125 countries with more than 90 percent of the world's population. For more than a quarter century he has advised dozens of heads of state and governments on economic strategy, in the Americas, Europe, Asia, Africa, and the Middle East. Prior to joining Columbia, Sachs spent over twenty years at Harvard University, most recently as Director of the Center for International Development and the Galen L. Stone Professor of International Trade. A native of Detroit, Michigan, Sachs received his BA, MA, and Ph.D. degrees at Harvard.

Laleema Senanayake is a researcher at LIRNEasia. Her research interests are regional development planning, rural connectivity and resilient communities. She has five years of experience in managing large-scale projects and conducting research locally and internationally. She currently manages a project in Nepal, which aim to facilitate and enrich policy discourse on improving broadband access by the poor and differently abled. Laleema is also involved in nationally representative research in Indonesia and Nepal to access Internet, mobile access and use and thereby developing comparable nationwide ICT indicators. She currently manages research in India and Myanmar to understand the opportunities for under-employed youth, women and previously excluded persons participating in micro-work and online freelancing platforms. Some of her previous research work focused on rural India, by assessing the potential of broadband networks for rural connectivity in India. Prior to joining LIRNEasia she worked as an intern in International Water Management Institute (IWMI) where she studied the small tanks and cascade systems of Sri Lanka. She also worked as a research assistant at University of Moratuwa to develop a climate resilient action plans for coastal urban areas of Sri Lanka. She holds a First Class Honours Bachelor of Science degree in Town and Country Planning. She has also passed qualifying examination of the Charted Membership of Institute of Town Planners Sri Lanka and is currently following her Masters in Economics at University of Colombo, Sri Lanka.

Hal J. Singer is a principal at Economists Incorporated, a senior fellow at George Washington's Institute for Public Policy, and an adjunct professor at Georgetown's McDonough School of Business. He has published several book chapters and his articles have appeared in dozens of legal and economic journals. Dr. Singer has testified before Congress on the interplay between antitrust and sector-specific regulation. His scholarship and testimony has been widely cited by courts and regulatory agencies. In several antitrust cases concerning class certification, the district court's order favorably cited Dr. Singer's testimony. In agency reports and orders, his writings have been cited by the Federal Communications Commission, the Federal Trade Commission, and the Department of Justice. Although his consulting experience spans several industries, Dr. Singer has particular expertise in the media industry. He recently advised the Canadian Competition Bureau on a large vertical merger in the cable television industry. He has served as consultant or testifying expert for several media companies, including Apple, AT&T, Bell Canada, Google, Mid-Atlantic Sports Network, NFL Network, Tennis Channel, and Verizon. Dr. Singer earned MA and Ph.D. degrees in economics from the Johns Hopkins University and a B.S. magna cum laude in economics from Tulane University.

Vincenzo Spiezia (Ph.D. in Economics) is the Head of the Information and Communication Technologies Unit in the Directorate for Science, Technology and Innovation of the OCDE. He coordinates the activities of the Working Party on Measurement and Analysis of the Digital Economy (MADE). His current research activities focus on the impact of ICT on employment, skills, and innovation. Before joining the OECD, he was Senior Economist at the International Labor Office in Geneva, where he contributed to a series of studies on the economic effects of globalization. He is author of several publications in books and international journals about innovation and employment. Vincenzo has a degree in Economic and Social Disciplines (DES) from the University Luigi Bocconi in Milan and a Ph.D. in Economics from the Italian Ministry of University, Research, Science and Technology.

Claudia Vittori is a post-doctoral fellow at the Department of Economics and Law of Sapienza University of Rome. In 2011, she was awarded a Ph.D. in Economics from University of Bristol. Her thesis endeavors the understanding of labor market inequality intragenerational mobility and polarization. She is also interested in social mobility. Two of her recent works examine intergenerational mobility in the UK in a lifetime perspective considering also people who have experienced workless spells. Alongside she is also working on a number of projects including examining the urban wage premium in Italy and the effects of labor market flexibility on innovation. Claudia has experience working with a range of data sources including the mature birth cohort studies (NCDS, BCS), the British Household Panel Survey (BHPS), the Survey on Income and Living Conditions (SILC), the Labor Force Survey and so on. Her research has been published in several international journals such as the *Oxford Bulletin of Economics and Statistics*, the *Economic Record*, the *Bulletin of Economic Research and Economia Politica*.

David Viviano is the Chief Economist at SAG-AFTRA (Screen Actors Guild—American Federation of Television & Radio Performers). In this capacity, David plays a primary role in negotiating and enforcing employment contracts for actors and other performers in television and film. David oversees SAG-AFTRA's Office of Media & Labor Economics, which analyzes evolving economic trends in television production and distribution. As a member of the senior executive staff at SAG-AFTRA, David collaborates with the elected leadership of the union to devise and implement strategy for the organization. Prior to serving as the Chief Economist of SAG-AFTRA, David served as the National Director of Research & Economics at the Screen Actors Guild. David earned a BA from Wesleyan University Majoring in Film Studies and holds an MBA. from Cornell University.

List of Figures

List of Tables

1

Introduction

Lorenzo Pupillo, Eli Noam and Leonard Waverman

The progress of technology in recent decades has been extraordinary. The great opportunities have been well recognised, but technology has also generated new divisions between winners and losers. In particular, information and communications technology (ICT) has been an engine of growth and transformation of economy and society but has impacted job flows and wage inequality. The result has been fear and uncertainty, and backlash. The purpose of our book is therefore to produce facts, analysis and evidence on the relationship between the diffusion of Internet and its impact on employment. The book also aims to fill a gap between academic research in this field, the cost and benefits of ICT diffusion and more general and accessible materials.

We took as our point of departure the structural transformation the world of work is currently undergoing. According to the Organization for Economic Cooperation and Development (OECD), three major forces are profoundly changing the world of work: demographic change, globalisation and technology, especially the digital revolution.[1] Among the demographic

L. Pupillo (✉)
Centre for European Policy Studies, Brussels, Belgium

E. Noam
Columbia Institute for Tele-Information, Columbia Business School, New York, USA

L. Waverman
DeGroote School of Business, McMaster University, Hamilton, Canada

© The Author(s) 2018
L. Pupillo et al. (eds.), *Digitized Labor*, https://doi.org/10.1007/978-3-319-78420-5_1

1

factors, the ageing population is capturing the attention of policymakers in the OECD countries for its impact on the affordability of health care and pension systems. These changes may suggest the need to create incentives for older workers to remain active in the labour market for a longer period of time. But such incentives may anger younger workers who feel that older workers are taking their jobs.

In the area of globalisation, the fragmentation of production processes and jobs along a global value chain is changing the occupational structure of jobs and their tasks. These powerful changes are often mistaken for the third force at work—the digitisation process. Increasing computer power coupled with the growing penetration of the Internet, Big Data, the Internet of Things (IoT) and Artificial Intelligence (AI) are profoundly changing the nature of work: by whom, where and how it will be performed.[2]

But what are the actual impacts of digitisation on labour? The current debate offers a variety of positions ranging from the pessimistic view of unprecedented job destruction, high rates of unemployment and massive increases in inequality to the optimistic idea that employment will adapt to the new technologies and that complementarities between humans and machines will generate new jobs and opportunities.

The most pessimistic study on the impact of digital technologies on future work is by Carl Benedikt Frey and Michael Osborn (2013). They evaluate the effect of what they call "computerization" on different occupations. They find that 47% of the total US employment can be classified in the "high-risk" category for becoming automated and therefore likely to disappear. A similar calculation for the UK finds a figure of one-third of jobs at high risk by automation over the next decades. However, their work has been criticised by a paper from the OECD (Arntz et al. 2016) claiming that occupation-based approaches tend to overestimate the impact of job automation, since occupations categorised as "high risk" may contain many tasks that are difficult to automate. Therefore, following the task-based approach instead of the broader occupational analysis, the OECD researchers' findings reduce the pool of jobs at high risk across the OECD countries to an average of only 9%.

A much more optimistic approach is suggested by the McKinsey Global Institute's report (2011), which concludes on the basis of its global SME survey that the Internet has created 2.6 jobs for every one job destroyed.[3] Enrico Moretti (2012) follows the path of digital technologies as powerful catalysts for job creation, and observes that high-tech jobs trigger a multiplier effect, increasing employment and salaries in non-high-tech sectors: one high-tech job creates five complementary non-high-tech jobs, both in skilled occupations (lawyers, teachers) and in unskilled ones (waiters, carpenters).

Beyond the issue of the net impact on employment, which our book deals with in several chapters, there are also fundamental questions related to the structural changes in the workplace that affect the nature of work itself. One of the most discussed issues is job and wage polarisation. As discussed in Chapter 3, a study by Autor et al. (2006) shows that the US labour market has become increasingly polarised both in terms of occupations and wage distribution between high-skilled, conceptual, non-routine jobs and low-skilled, manual, but non-routine jobs at the opposite end of the jobs/income spectrum. The jobs most at risk from IT, however, are the routine, middle-skilled jobs, set in the middle of the jobs or wage distribution. This process is explained by the hypothesis that computers substitute for workers in carrying simple cognitive and manual activities ("routine" tasks), while computers complement workers in carrying out problem-solving and complex communication activities ("non-routine" tasks). Goos et al. (2009) show these findings for Europe. The jobs and wages polarisation has determined what was has been called the great "hollowing out" of the middle class: higher paying jobs requiring high skills and creativity have proliferated; demand for low skills, manual jobs not replaceable by computers has also increased; but middle skills, middle-class jobs like bookkeeping, clerical work and routine manufacturing jobs, have been substituted by computers and the Internet. These processes combined with some of the effects of globalisation also have strong socio-economic implications, which have contributed to recent seminal political outcomes such as the Brexit vote in the UK, Trump's election in the United States and the rise of populism in Europe.

This labour market process of substitution of computers for workers to execute routine jobs, called routinisation and often associated with labour market polarisation, seems to signal the deeper and more structural effects of how digitalisation will affect our lives. According to Brynjolfsson and McAfee (2011, 2014), the technological progress is "at an inflection point – the early stages of a shift as profound as that brought on by the Industrial Revolution".[4] This transformation brings a bounty of innovation related to the exponential, digital and combinational nature of the new digital technologies. While we have experienced so far only the transformation of the way in which we communicate, most of the gains from the digital transformation are still ahead of us. However, with the bounty brought by these technologies also comes the tendency to replace human capital in skilled tasks once considered safe from automation.

But is this not a scenario we have already witnessed over the past two centuries? Technological innovations have always disrupted the job market, but after a period of labour turmoil, previous technological revolutions have brought more jobs and prosperity. The question now is: *Is it different this time and if so, why is that the case?* This question resides at the centre of Nigel M. de S. Cameron's 2017 book *"Will robots take your job?"*. Two dimensions may make it reasonable to think that this wave of technological disruption could, in fact, be different: *time and scale*.

We are in the midst of a transition to a digital economy, and there is widespread consensus that the transformation brought by the digital technologies in the economy and society today is proceeding *at much faster speed* than those generated by previous waves of technologies.[5] Although this digital transformation started in the middle of the last century, the pace of change has accelerated as digital infrastructure is further deployed and powerful devices like smartphones connect people everywhere at any time.[6] Tom Standage, digital editor for *The Economist*, interviewed for a Pew Research on the future of jobs stated:

> Previous technological revolutions happened much more slowly, so people had longer to retrain, and [also] moved people from one kind of unskilled work to another. Robots and AI threaten to make even some kinds of skilled work obsolete (e.g., legal clerks). This will displace people into service roles, and the income gap between skilled workers whose jobs cannot be automated and everyone else will widen. This is a recipe for instability.[7]

The McKinsey Global Institute states that AI is contributing to a transformation of society "happening ten times faster and at 300 times the scale, or roughly 3000 times the impact" of the Industrial Revolution.[8] The speed of this technological change is linked to the exponential trends in technology development and its evolution may be faster than our reaction capacity. According to some observers, the acceleration of computer hardware over decades, fuelled by Moore's Law, has allowed us to "remain on the steep part of the S-curve for far longer than has been possible in other spheres of technology".[9] All this means that it is difficult to predict what will happen and that the changes in the future will likely be massive.

The breadth of change is major and the impact of technological change is more pervasive than past changes that were linked to the diffusion of other technologies. Their impact is much more widespread than in the past. All sectors of the economy are involved. The example of self-driving vehicles is a case in point. Not only will many driving jobs disappear (truck, bus and taxi drivers), but also with fewer vehicles purchased there will be an impact on car manufacturing, the petroleum industry, farming (ethanol) and the auto insurance industry. Greater safety will also have an impact on the health care industry. Furthermore, these effects will not be geographically limited as was the case in the closing of factories in steel towns or mines in coal-mining villages.[10]

A helpful contribution to the discussion of the impact of automation on jobs comes from David Autor (2014) who draws our attention to the possible complementarity between people and machines. He starts by recalling the work of the Hungarian scientist and philosopher Michael Polanyi who in 1966 observed: "We can know more than we can tell…The skill of a driver cannot be replaced by a thorough schooling in the theory of a motorcar; the knowledge I have of my own body differs altogether from the knowledge of its physiology".[11] This statement, also known as the Polanyi's paradox, emphasises the difference between tacit knowledge and explicit knowledge and characterises the interplay between humans and computers. The computers are quite good at replicating explicit, codifiable procedures such as multiplication. However, tasks that require flexibility, judgment and common sense, skills that we understand only tacitly, are poorly performed by computers. In other words, humans have comparative advantages when flexibility, problem-solving skills and creativity are required. Autor observes that the extent to which a machine can substitute for human labour has been overstated while the strong complementarities between computers and human labour that, increase productivity and demand for skilled labour, have been ignored. A task that cannot be computerised does not mean that computerisation has no effect on that task. On the contrary, tasks that cannot be computerised are generally complemented by the technology. Think about the role played by mechanisation in construction. Using cranes, excavators and pneumatic nail guns, the amount of physical work that a worker can accomplish in an eight-hour work day increased tremendously. Of course, automation has substituted for human labour in construction but has not devalued the work of construction workers. Indeed, no matter how much capital equipment is available, a construction site without construction workers would be useless. Construction workers perform ad hoc tasks such as control, guidance and flexibility in managing everyday situations. Therefore, automation has complemented construction workers. Yet, technology is going further and deeper.

The idea of the complementarity between humans and machines, such as through highly capable robots, needs to be considered dynamically, taking into account how good machines are becoming in substituting for workers in functions that require flexibility, once considered a unique characteristic of the human being. The recent victory of ALPHAGO, the artificial intelligence system built by Google–DeepMind, over the human champion

Lee Se-dol in the strategy game of Go, shows the power of the combination of deep learning systems with reinforcement learning techniques. This combination is making viable a new approach in gaming. Instead of trying to incorporate smart strategies in the computer, this new approach allows the computerised system to build and learn winning strategies almost by itself.[12] For how long thus will the Polanyi's paradox hold? According to a recent book by McAfee and Bryjolsson (2017), "computers still don't really understand the human condition […] and digital technologies do a poor job of satisfying most of our social drives. So, work that taps into these drives will likely continue to be done by people for some time to come. Such work includes tasks that require empathy, leadership, teamwork, and coaching".[13]

Against this challenging background, our book attempts to contribute to the debate on the impact of the Internet on jobs and to complement the existing literature. It is first necessary to clarify that the chapters presented in the book deal primarily with the impact that computers and networks are having on jobs, (the "third industrial revolution") rather than on robots, artificial Intelligence, the Internet of Things or Big Data. The nature of broadband or Internet as a general-purpose technology makes it difficult to disentangle the specific contribution of application technologies. The empirical chapters presented in the book focus on the fundamental building blocks rather than more specific refinements and applications of these technologies, while the more theoretical chapters discuss these issues in broader terms.

The first chapter in our book confirms the economists' view that the displacement of jobs due to the diffusion of computers and networks is a short-term issue and that employment will not be reduced in the long run. The chapter shows that ICT investments have no effects on labour demand in the long run. More precisely, it claims that the substitution effect and the scale effect compensate each other completely and that the short-run disruption effects on the job market tend to disappear in about 20 years in most OECD countries.

The book also provides new quantitative evidence of the impact of the Internet on jobs. On the issue of routinisation, focusing on this process in Europe, the book shows clearly that individuals in routine jobs, ceteris paribus, display a higher probability to become unemployed, challenging the idea of recent decades that only unskilled workers represent the vulnerable segment in the labour market. All the empirical chapters indicate that the Internet has a positive impact on the creation of jobs, but it is not clear if the jobs created outnumber the ones destroyed. In the media and entertainment business, for instance, the Internet has significantly changed the labour market for talent. It has added new jobs for performers, but these new jobs tend to be at the lower end of the pay spectrum. Overall, it is unclear whether the incremental gains in lower-paying jobs are meaningful enough to outweigh the increased competition that characterises the life of professional performers today.

Looking at the impact of Internet connectivity on employment, the book shows that broadband contributes to the creation of jobs in certain industries and geographies. However, broadband is also a key factor in capital-labour substitution in sectors such as tourism or in rural areas where, contrary to what happens in metropolitan areas, job losses driven by productivity enhancements cannot be compensated by innovation-led new business models.

We also present an ad hoc approach to the estimation of the balance between gains and losses of employment in the digital economy. It uses a bottom-up accounting of both job destruction and job creation. While the effects vary significantly by location and sector, the overall outcome is a small but constant growth in most places where investments in digital economy are robust.

Besides offering an assessment of the quantitative impact of the Internet on jobs, chapters in the book also discuss the distributional effects of digital technologies on income inequality and insecurity, and the impact on the middle class in western countries. The first concern raised is linked to the unequal distribution of job losses. The loss is mostly concentrated in low-pay industries, and job mobility from the lower to middle-class is slowing. There is also a generational inequality created by the Internet: older people become obsolete faster than before. Is this a global phenomenon? The book discusses the case of Canada and suggests that wage and jobs polarisation are not as present as they are in the United States. Is this related to the fact that Canada lags the United States in many types of ICT infrastructures, application adoption and usage? It is too early to tell. As is the case for other technological revolutions, it takes time for change to fully manifest its effects.

We also point out in the book the potential contribution to the economy that may come from the structural change in US infrastructure from a rigid hardware-based architecture to a software-based infrastructure, which is flexible and fluid in topology and capable of scaling clients and resources on demand.

All contributions in the book point to the need to govern the changes in the labour market induced by the Internet through new approaches and solutions. These changes in the labour market should not be managed on a "business as usual" basis since there is nothing routine about structural unemployment. The speed and the scale of changes induced by the digital revolution require new education and labour market adjustment policies that facilitate structural and social adjustment without slowing innovation.

There is a strong need for a massive improvement of ICT skills of the population at large. Such skills are needed, not just in the ICT sectors or in closely related finance/insurance, professional services and real estate. A lack of ICT skills threatens the growth and well-being of the economy more generally. Chapters in the book show how the private sector is responding to these challenges by retraining its labour force. Also, the private sector is providing support for general improvement in secondary and high school education and post-secondary science, technology, engineering and math (STEM) career skills. One chapter suggests that this process should be encouraged by government through a "knowledge tax credit" by allowing qualified expenditures on R&D and workforce training to offset taxes. The author of that chapter also urges governments to ensure that education is more closely linked to occupational needs, such as in nano- and micro-technology, rapid prototyping, logistics and bio-manufacturing.

Furthermore, special attention should be devoted to help workers who lose their jobs, through retraining and subsidies, without reducing the incentive for workers to return to the workforce. A chapter in the book discusses the role that fiscal policy can play in making robotic innovation a win-win development for everyone. More generally, policies must avoid a "crowding out" of private investment when subsidising the diffusion of innovation, such as broadband networks.

Yet another role for online platforms is that of job matching and transforming informal work into formal employment, especially for emerging economies.

1.1 Key Messages from Our Book, Chapter by Chapter

The chapters of the book are grouped thematically into three parts: (1) The Impact of Technological Change on Jobs; (2) Internet Economic Fundamentals and their Impact on the Economy and Distribution; and (3) Policies to Facilitate Structural and Social Adjustments without Slowing Innovation. In the following sections, we highlight the main ideas contained in the book's chapters.

1.1.1 The Impact of Technological Change on Jobs

In Chapter 2, entitled, "ICT Investments and Labor Demand in OECD Countries", Vincenzo Spiezia, Senior Economist at the OECD, provides new estimates of the effect of ICT investments on labour demand in 19 OECD countries over the period from the early 1990s to 2012. He measures ICT technical progress as the decline in the user cost of ICT capital and estimates the effects of such a decline on the demand for labour. The findings suggest that ICT investments have no effects on labour demand in the long run. A permanent decrease in the user cost of ICT capital reduces labour demand per unit of output but it increases output by the same proportion. In other words, the substitution effect and the scale effect compensate each other completely. In the short run, however, due to sluggish adjustments in production inputs, a one-off permanent decrease in ICT user cost results in a temporary increase in labour demand followed by a temporary decrease. Spiezia's estimates suggest that these temporary effects tend to disappear in about 20 years in most OECD countries. While the negative employment effects of ICTs are estimated to eventually fade away, their medium-term persistence justifies policy measures such as investment incentives, labour market activation policies and temporary income support.

In the third chapter of Part I, Jeffrey Sachs of Columbia University and Seth Benzell and Guillermo LaGarda, of Boston University, approach the issue of whether robots raise or lower economic well-being. In their chapter entitled "A One-Sector Model of Robotic Immiserization", they claim, on the one hand, that robots raise output and bring more goods and services within consumers' reach. On the other hand, they eliminate jobs, shift investments away from machines that complement labour, reduce wages and harm workers who cannot compete. The net effect of these offsetting forces is unclear. In a one-sector model, the authors investigate a possible mechanism by which innovation in automation may reduce welfare. They show in this setting how fiscal policy can make robotic innovation a win-win for all generations. Indeed, policies that redistribute income across generations can ensure that a rise in robotic productivity benefits all generations.

In the fourth chapter, entitled "Routinization and the Labor Market: Evidence from European Countries", a group of Italian economists— Federico Biagi, University of Padova, Paolo Naticchioni, Roma TRE University, Giuseppe Ragusa, LUISS University, Claudia Vittori, University of Rome La Sapienza—provide new evidence on the ongoing routinisation processes in Europe. In particular, they claim that routinisation processes do not depend on the type of variable considered; groups of countries characterised by very different institutions, cultures and labour market conditions share similar routinisation trends; routinisation levels are different across groups, suggesting that groups of countries can be placed at different stages in the technological/routinisation process; routinisation seems to represent a driver, among others, of unemployment inflows and finally individuals in routine jobs, ceteris paribus, have a higher probability of becoming unemployed. This evidence challenges the view of earlier decades that only unskilled workers represent the weak segment in the labour market. This evidence applies also to routine workers, who are not necessarily unskilled. If confirmed, this issue would represent an important new dimension in policy debates in the coming years, to think about some new targeting dimensions of the active and passive labour market policies, with two main objectives. On the one hand, such policies may ease the reallocation of routine workers towards other types of jobs, and on the other hand, they may provide some sort of additional institutional insurance for workers exposed to higher unemployment risks.

The fifth chapter, "Labor Markets in the Digital Economy: Modeling employment from the bottom-up" by Jonathan Liebenau from the London School of Economics, claims that the effects upon labour markets of changes in the digital economy are poorly understood because analyses have not focused on the appropriate level of jobs within specified sectors. Liebenau, instead, uses a bottom-up accounting of both job destruction and job creation that provides a more accurate assessment of the effects of investment in the digital economy. While the effect differs significantly from sector to sector and place to place, overall he observes modest but steady growth in most places where investments in the digital economy are robust. This is an important finding both for the general understanding of the effects of the digital economy and for our ability to be specific about the impacts of investments of certain kinds on the local economies. It provides a clearer idea of labour market skills requirements and this helps monitor education and training policies, if only to evaluate the refrain coming from employers and amplified by researchers, who perennially claim that the labour market is desperately in need of hundreds of thousands of newly skilled workers.

In particular, the studies show some surprising results. Trade data and industry structure evidence indicate that the UK economy is more capable of meeting the expectations of digital economy requirements in areas such as smart metering and intelligent transport systems infrastructure than might have been expected. Their reliance on imports may be less than sometimes assumed and fewer of the jobs generated by such spending will be created abroad. Another surprise finding is the strong effect that policies regarding restricting the export of data and on energy pricing have on the ability of cloud services to generate domestic employment. Especially given that cloud service provision is highly portable, the ability to avoid exporting all new employment opportunities rests on the local economy's ability to achieve growth in those sectors that benefit from the new digital services, as well as to attract the providers of those services who rely upon data centres and skilled staff.

In Chapter 6, "The Impact of the Broadband Internet on Employment", Raul Katz from Columbia University states that the topic of the impact of the broadband Internet on employment has been present in the public policy arena on a regular basis in recent years. Unfortunately, such an important debate has been approached with little formalisation of the research of impact and understanding of the evidence. Before even tackling the prescriptive side of the policy debate, researchers appear to be aligned in one of two camps: either that the Internet contributes to the creation of jobs; or, that the Internet is the source of job destruction. Unfortunately, in many cases, research is being conducted hypothetically (e.g. what kind of jobs are susceptible to elimination as a result of digitisation?) without looking at the empirical evidence. The chapter summarises the results of investigations conducted by this author and other researchers on the impact of broadband on employment. This chapter argues that, based on the evidence, the response to the question of impact of broadband on employment is: it all depends. The chapter shows that broadband contributes to the creation of jobs in certain industries and geographical areas, while also being a key factor in capital-labour substitution (i.e. job losses) under certain conditions. This kind of differentiated answer does not satisfy pundits or ideologues. In the final analysis, however, if policymakers are oriented towards making good decisions, they need to have a solid, unbiased understanding of the evidence.

In the final chapter of Part I, "The Impact of the Internet on Employment and Income in the US Media and Entertainment Business", David Viviano, chief economist at the SAG-AFTRA Labor Union of actors and other performers, emphasises that the Internet has undoubtedly changed the labour market for talent in the media and entertainment business. The Internet's impact on the sector is still evolving, but some insights can already be drawn. The Internet has added new jobs for performers; however, the vast majority of these new jobs tend to be at the lower end of the pay spectrum. The Internet has also created new ways to distribute and monetise library video content. While these new revenue streams are becoming more significant in the aggregate, they have yet to become a significant source of income for individual performers. Furthermore, the Internet has disrupted this labour market by driving up competition and increasing year-to-year income variability. At this point, it is unclear whether the incremental gains in lower-paying jobs and new but smaller "residuals" payments to actors are high enough to outweigh the increased competition, wage compression and income variability that professional performers face in the Internet age.

1.1.2 Internet Economic Fundamentals and Their Impact on the Economy and Distribution

The first chapter in Part II (Chapter 8), by Eli Noam from Columbia University—"Inequality and the Digital Economy"—assesses the distributional effects of the Digital Economy. Noam opens with the observation that policymakers in developed countries have believed and hoped for many years that the Internet, and more generally the digital economy, would replace and enhance industrial jobs. Yet after several decades of digital evolution, one can observe that the Internet has displaced many blue-collar jobs in manufacturing as well as pink-collar jobs in retailing and among clerical staff. It has also created new jobs, but those who get the new jobs are not the same people who have lost them. Furthermore, what is striking is not the loss per se but the fact that the losses have not been distributed equally. A majority of the jobs were in low-paying industries and the traditional job mobility from lower- to middle-class is becoming more difficult. There is also a generational inequality accelerated by the Internet: the rapid transformation in knowledge and technology makes the experience less valuable and quickly renders the old out-of-date. It is necessary to realise that the impact of Internet-induced economic displacements in developed economies will not go away. It will get worse in the short term. This creates a challenge to managers and policymakers in the digital sector. Otherwise, a backlash will create forces that will restrict innovation. It is therefore important for academics, public-policy analysts, NGOs, companies and governments to think creatively about new approaches to these issues, and to balance the public interest, technological innovation and financial investment in the emerging environment.

In Chapter 9, "Job Losses and the Middle Class: Canada and the USA, and the Possible Role of ICT", Len Waverman, of the DeGroote School of Business, McMaster University uses comparisons between Canada and the United States to examine whether recent job losses in the middle-class is a general phenomenon, or more uniquely an American experience. The role of ICT in such "polarisation" is examined and the Canadian productivity and ICT experience relative to the United States are discussed. While data indicate that wage and job polarisation have not occurred in Canada as of 2013, Professor Waverman claims that "it's too early to tell".

The explosion of social media, viral networks and applications spawned in 2007 by the iPhone are now only a few years old! Most advances have been directed at the consumer market. Even there, new advances such as self-driving cars and virtual reality are not market ready. We have not yet begun to tap the enormous potential. It is important to remember that other general-purpose inventions were slow to come to fruition and uneven in timing and impact. Productivity will improve as the ICT revolution continues to expand beyond consumer-driven social media. Yet, the US data show that ICT, at least in the short to medium term, has been a source of growing income disparity because of the displacement of "routine" jobs. The level of skill required to move from routine-based jobs to non-routine or cognitive occupations is high. These are very different transitions from the past. Rich countries—the West—have done little to enable such job/occupational shifts. And for those who cannot at their life cycle stage make such adjustments, we have not created the necessary social safety nets.

In the final chapter of Part II, "Internet Innovations—Software Is Eating the World: Software Defined Ecosystems and the Related Innovations Result in a Programmable Enterprise", Robert Cohen, senior economist at the Economic Strategy Institute, claims that a structural change in US infrastructure from a rigid, inflexible, hardware-based architecture ("The Old IP") to the "New IP"—a software-based infrastructure that is flexible, fluid in topology and architecture, and capable of scaling clients and resources on demand—is underway. This makes it possible for US firms that are early adopters to expand the optimisation of their current operations. Cohen estimates that these firms are likely to add $3 trillion to the US economy over the next 10 years and about 8 million new jobs. His essay describes the software-based innovations that are changing the Internet and creating new economic opportunities. These innovations have been well documented as individual technology advances. Here, the author links them together to argue that as the Internet evolves to a new stage, it is providing a series of changes that, operating in concert, create new environments for businesses and consumers.

1.1.3 Policies to Facilitate Structural and Social Adjustments Without Slowing Innovation

Chapter 11, which opens Part III of this book, is entitled "ICT Innovation, Productivity, and Labor Market Adjustment Policy", by Robert D. Atkinson, of the Information Technology and Innovation Foundation. He states that there is increased interest in the question of how to facilitate labour market adjustments from technological innovation. Some of this interest appears to be a result of efforts to try to respond to the recent increase in populism from both the right and the left, fuelled, as some believe, by labour market insecurities. Some are due to the weak global labour market performance in the wake of the Great Recession, where tens of millions of jobs were destroyed, and job creation has been tepid. And some is due to the belief that technological change, particularly in the information and communications technology sector is fuelling (or about to fuel) rapid productivity growth and accompanying job loss. Regardless of the reasons, improving worker adjustment policies in the United States is long overdue. The alternatives—doing relatively little—risks not only increasing opposition to ICT-driven technological change but reducing the efficiency of the labour market. He concludes that the major risk to the global economy over the next decade is not too much disruption, but too little. In other words, the risk is that productivity will grow too slowly. As such it is critical that labour market policies, including adjustment policies, support, not hinder ICT-led creative disruption. One way to do that is to make improvements in workforce training and labour market adjustment policies.

Chapter 12, "Ensuring the Education and Skills Needed for ICT Employment and Economic Growth" by Richard Clarke from AT&T, offers a view from the private sector on the issue of mitigation and adjustment policies. Clark points out that over the last 40 years, ICTs have been the most important source of economic growth and sectoral change in the output and labour markets in America. Indeed, it is hard for anyone under the age of 50 to imagine a time when manual processes were the exclusive way of conducting a financial transaction or making an airline reservation. But with these changes in technology have come great changes in the educational requisites for workers. STEM knowledge and advanced education are now the key to productive employment and a middle-class lifestyle. While traditional educational systems have faced difficulties in effectively providing widespread STEM education, ICT organisations themselves may offer a partial remedy. By harnessing the ability of ICTs to convey such education via online MOOCs or nanodegrees, it may be possible to ensure that the economic revolution that ICTs have wrought will continue into the twenty-first century.

An additional view from the private sector comes in Chapter 13 by Andrea Iapichino, Amelia De Rosa and Paola Liberace, all from Telecom Italia (TIM) and entitled "Smart Organizations, New Skills and Smart Working to Manage Companies' Digital Transformation". The authors discuss the implications of digital transformation for organisations and the future of work. They first describe the drivers of digital transformation, both social and technological, and focusing on the profile of the new "smart organization" and its required skills. They then present TIM's way to enact the transition from knowledge to skills, through its TIM Academy initiative and Knowledge Management process, and towards a new way of working through its Smart Working project. The digital transformation can be a great opportunity for the Company if this process is managed throughout the range of peoples' skills and engagement.

A contribution from the developing world is Chapter 14 "Investigating the Potential for Micro-Work and Online Freelancing in Sri Lanka", by Helani Galpaya, Suthaharan Perampalam, Laleema Senanayake, of LIRNEasia. Sri Lanka is a lower middle-income country that has benefited significantly from the increase in business process management (BPM) work from developed countries. Many global BPM operations have set up business in Sri Lanka and provide a range of services to overseas clients. While there is much room for improvement in computer literacy and Internet access (in 2015, 26.8% of the population was computer literate and over 25% of households had a computer), there is wide enough diffusion of computers to make participation in micro-work a viable option for many, beyond the educated elite. Therefore, using focus groups and a sample survey, the chapter asks whether it is possible that the digital dividend could be spread more inclusively through the participation on micro-work platforms. And if so, how do we encourage this? The authors show that whether people engage in full-time or part-time work, online work presents a way to increase their income, sometimes significantly. There is a sufficient number of people in the country with the right type of skills exist, but it has to be "pitched" in the right way. Awareness has to be increased, including a recognition of the pitfalls (e.g. not getting paid for work done), and the solutions to these pitfalls. Policymakers need to facilitate a legitimate payment mechanism, which is an oft-cited problem. This chapter concludes that online freelancing and microwork presents a growth opportunity for Sri Lanka.

The final contribution of the book (Chapter 15), entitled "Do Municipal Broadband Networks Stimulate or Crowd Out Private Investment? An Empirical Analysis of Employment Effects", by Hal Singer, senior economist at Economics Incorporated, addresses the role of public funding for broadband networks at the municipal level. Economists have studied the employment and spillover effects from broadband investment generally, as well as the employment effects from municipal-owned broadband networks. In contrast to the well-documented link between broadband deployment and jobs, the literature finds scant evidence that muni-networks are associated with private-sector employment gains. Economic theory offers a possible explanation for this result—namely, private firms do not wish to compete against government-owned enterprises, which do not take profits into consideration when setting prices. Accordingly, any incremental investments by local firms wishing to exploit the muni-network could be offset by forgone investments by privately owned Internet service providers (ISPs). NTIA data of US broadband deployment by both privately and muni-networks in December 2010 tends to corroborate this "crowding-out hypothesis". US counties that enjoyed privately owned broadband deployment in 2010 realised faster-than-average private-sector employment growth from 2010 through 2013, whereas US counties that relied on government provision of high-speed networks in 2010 experienced no similar lift in private-sector employment growth. This chapter concludes by spelling out various policy implications of these findings.

1.2   Conclusion

Our book brings new evidence of the impact of the Internet on jobs. All of the empirical contributions indicate that the Internet has a positive impact on the creation of jobs, but it is not clear if more jobs are created than are destroyed. Furthermore, routinisation, job market polarisation and new labour market inequalities show that, while the diffusion of the Internet is generating opportunities for many, it is also accompanied by ambiguous trends that by themselves will not generate a more resilient and inclusive labour market.

The book points to the need to manage these changes in the labour market neither as "business as usual", since there is nothing routine about structural unemployment, nor the rapid speed at which changes are occurring. Failing to mitigate "short-term" job losses risks a policy response to reduce both the speed and the extent of the ICT revolution. Twenty years is not the short run for those affected... It is a generation! Technological innovation has always had disruptive effects on the job market. To ignore the "short run" is to risk bringing back a modern version of the Luddites, with profoundly negative consequences.

Some chapters in the book refer to the new efforts required in retraining the labour force and in STEM career skills acquisition and improvement. But one area of particular interest that needs further research is the analysis and design of the new set of skills that allow working alongside the smart new machines. This means researching how to better understand the complementarity between man and machine and how to develop the skills that allow men and women to do "what computers are not good at".[14] Moving from STEM (science, technology, engineering and mathematics) to STEAM (adding arts to the mix) would be a right step in this direction. Furthermore, soft skills such as leadership, team-building and creativity will become increasingly important and are less likely to be automated.

Thus, the future of work could be seen not in its replacement or displacement by technology but in the complementarity between humans and machines. Smarter machines and smarter people can complement each other to create a mass of customised products and services: the world of new artisans.[15] In the era of the Paradox of Progress, it is up to us to work in a way that allows the promise to prevail. This book is a small effort in this direction.

Notes

1. See Scarpetta, Stefano (2016), "What Future for Work?" *OECD Observer*, No. 305, Q1.
2. Ibid.
3. Matthieu Pelissie du Rausas et al. (2011), *Internet Matters: The Net's Sweeping Impact on Growth, Jobs and Prosperity*, McKinsey Global Institute.
4. Brynjolfsson, Erik, and Andre McAfee (2014), *The Second Machine Age*, New York, NY: W. W. Norton, p. 251.
5. See, Manyika, James (2017), *Technology, Jobs and the Future of Work*, McKinsey Global Institute, May.
6. Note that the World Wide Web began in 1996 and the iPhone debuted on 29 June 2007!
7. Smith, Aaron, and Janna Anderson (2014), "AI, Robotics and the Future of Jobs," Pew Research Center, August, p. 11.
8. *The Economist* (2016), "The Return of the Machinery Question," June 25.
9. Ford, Martin (2015), *Rise of the Robots*, New York, NY: Basic Books, p. 69.
10. See De S. Cameron (2017), op. cit., p. 69.
11. Autor, David (2014), "Polanyi's Paradox and the Shape of Employment Growth," NBER Working Paper No. 20485, National Bureau of Economic Research, Cambridge, MA. This paragraph draws from this paper.
12. McAfee, Andrew, and Erik Brynjolsson (2016), "Where Computers Defeat Humans, and Where They Can't," *New York Times*, March 16.
13. Andrew McAfee and Erik Brynjolfsson (2017), *Harnessing Our Digital Future—Machine, Platform, Crowd*, New York, NY: W.W. Norton.
14. See McAfee and Brynjolfsson (2017).
15. See Rework America (2015), *America's Moment*, New York, NY: W.W. Norton, p. 131.

References

Arntz, Melanie, Terry Gregory, and Uldrich Zierahn. 2016. "The Risk of Automation for Jobs in OECD Countries: A Comparative Analysis." OECD Social, Employment and Migration Working Papers, OECD, Paris, May 14.

Autor, David. 2014. "Polanyi's Paradox and the Shape of Employment Growth." NBER Working Paper No. 20485, National Bureau of Economic Research, Cambridge, MA.

Autor, David, Lawrence Katz, and Melissa Kearney. 2006. "The Polarization of U.S. Labor Market." *American Economic Review* 96 (2): 184–194.

Benedikt, Frey Carl, and Michael A. Osborn. 2013. *The Future of Employment: How Susceptible Are Jobs to Computerization?* Oxford: Oxford Martin School.

Brynjolfsson, Erik, and Andrew McAfee. 2011. *Race Against the Machine: How the Digital Revolution Is Accelerating Innovation, Driving Productivity, and Irreversibly Transforming Employment and the Economy.* Lexington, MA: Digital Frontier Press.

Brynjolfsson, Erik, and Andrew McAfee. 2014. *The Second Machine Age: Work, Progress, and Prosperity in a Time of Brilliant Technologies.* New York, NY: W. W. Norton.

De S. Cameron, Nigel M. 2017. *Will Robots Take Your Job?* Cambridge, UK: Polity Press.

du Rausas, Matthieu Pelissie, James Manyika, Eric Hazan, Jacques Bughin, Michael Chui, and Rémi Said. 2011. *Internet Matters: The Net's Sweeping Impact on Growth, Jobs and Prosperity.* New York, NY: McKinsey Global Institute.

Ford, Martin. 2015. *Rise of the Robots: Technology and the Threat of a Jobless Future.* New York, NY: Basic Books.

Goos, Maarten, Alan Manning, and Anna Salomons. 2009. "Job Polarization in Europe." *American Economic Review* 99 (2): 58–63.

Manyika, James. 2017. *Technology, Jobs and the Future of Work.* New York, NY: McKinsey Global Institute.

McAfee, Andrew, and Erik Brynjolfsson. 2016. "Where Computers Defeat Humans, and Where They Can't." *The New York Times*, March 16.

McAfee, Andrew, and Erik Brynjolfsson. 2017. *Machine, Platform, Crowd—Harnessing Our Digital Future.* New York, NY: W. W. Norton.

Moretti, Enrico. 2012. *The New Geography of Jobs.* New York, NY: Houghton Mifflin Harcourt.

National Intelligence Council. 2017. *Global Trends. Paradox of Progress*, January. Washington, DC.

Smith, Aaron, and Janna Anderson. 2014. "AI, Robotics and the Future of Jobs." Pew Research Center, August.

Rework America. 2015. *America's Moment.* New York, NY: W.W. Norton.

Scarpetta, Stefano. 2016. "What Future for Work?" *OECD Observer*, No. 305, Q1.

The Economist. 2016. "The Return of the Machinery Question," June 25.

Part I

The Impact of Technological Change on Jobs

2

ICT Investments and Labour Demand in OECD Countries

Vincenzo Spiezia

2.1 Introduction

Major technological innovations in economic history have always been accompanied by major transformations in the labour market. By increasing labour productivity, innovation enables producing a given amount of goods and services with less employment, thus leading to the possibility of *technological unemployment*. At the same time, innovation triggers a number of *compensation mechanisms* with potential positive effects on employment.

Information and communication technologies (ICTs) are no exception to this historical pattern. Information technologies replace workers that perform routine tasks with computer-directed production processes (*automation*). Furthermore, communication technologies allow coordination of complex production activities across space and delocalisation of labour-intensive productions activities to low-wage countries (*offshoring*). At the same time, ICTs create new employment opportunities in the ICT sector and in the whole economy.

The overall effect of these different factors is predicted to be positive under the conditions postulated by economic theory in the long run. As economies may deviate from these conditions in the short run, the net employment effect of ICTs is likely to depend on institutions and policies.

V. Spiezia (✉)
OECD, Paris, France

© The Author(s) 2018
L. Pupillo et al. (eds.), *Digitized Labor*, https://doi.org/10.1007/978-3-319-78420-5_2

This chapter provides new estimates of the effects of ICT investments on total labour demand in 19 Organization for Economic Cooperation and Development (OECD) countries over the early 1990s–2012. By looking at the total economy, these estimates enable measurement of both the positive and negative employment effects of ICTs, which recent studies at the firm or industry level cannot account for.

The chapter is organised as follows. Section 2.2 summarises the main predictions of the economic theory on the effects of innovation on employment while Sect. 2.3 reviews recent empirical studies on the topic. Sections 2.4 and 2.5 introduce the model and the data for the analysis. The main findings are discussed in Sect. 2.6 while Sect. 2.7 concludes.

2.2 ICTs and Employment: What Does Economic Theory Say?

The analysis of the effects of innovation on employment goes through the history of modern economics, e.g. Say, Ricardo, Marx, Hicks, Marshall and Keynes, among others. The results of this analysis are known in the economic literature as "compensation theory." At the core of compensation theory is the prediction that, while innovation may reduce labour demand and lead to unemployment, it also triggers a number of automatic mechanisms that are expected to compensate for the direct decrease in labour demand. The compensation theory provides useful insights on the effects of ICTs on employment (OECD 1994; Spiezia and Vivarelli 2002).

Figure 2.1 provides an illustration of the opposing forces at play. Changes in employment (L) are the results of growth in output (Y) and the changes

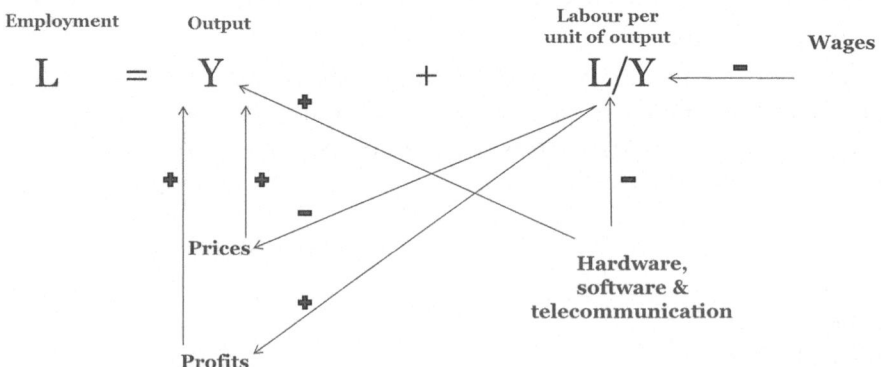

Fig. 2.1 Effects of ICTs on employment

in the quantity of labour required to produce one unit of output (L/Y). As shown in the figure, ICTs have an impact both on labour requirements and on output.

To start with, it is useful to distinguish between process innovations and product innovations. A process innovation increases productivity and reduces unit costs whereas a product innovation results in the commercialisation of new goods and services. Interestingly, ICTs comprise both process innovations, e.g. computer-controlled machineries, automated inventory flows and product innovations, e.g. smartphones, e-books, etc.

By increasing total productivity, ICT process innovations enable firms to produce a given amount of goods and services with less employment, thus leading to the possibility of technological unemployment. This effect is stronger the larger the *labour-saving bias*[1] of the new technology, i.e. the more ICTs reduce the demand of labour relative to that of capital, at constant input prices. The labour-saving bias may be different depending on the type of labour, e.g. ICTs tend to be biased against low-skill workers and towards high-skill labour.

At the same time, ICT process innovations lead to lower unit costs of production. In a competitive market, this decrease is translated into lower prices, which stimulate higher demand for products. In turn, higher demand generates additional production and employment (compensation "via decrease in prices"). The strength of this effect depends positively on two factors: first, the degree of competition in the product markets; and, second, the price elasticity of final demand.

In less competitive product markets, the decrease in unit costs induced by ICTs is not fully translated into prices and generates extra-profits for the innovative firms. Part of these extra-profits is directly reinvested and increase production and employment in the capital good sector (compensation "via increase in machineries"). The other part provides additional income for share-holders (as dividends) and workers (through wage bargaining), who may spend it on higher consumption or save it. Higher consumption directly increases aggregate demand while savings are lent through the financial system to finance investment by firms and consumption by households. Through these different channels, the increase in income generated by ICTs raises aggregate demand, production and employment (compensation "via increase in income"). The strength of these effects would be larger the higher the firms' propensity to invest, the higher the households' propensity to consume and the higher the efficiency of the financial system to reallocate savings.

The direct effect of ICT process innovation on employment may be further compensated by a decrease in real wages, which leads to an increase in the labour intensity of production and/or to a decrease in unit production costs (compensation "via decrease in wages"). The strength of the former effect depends on, first, the degree of substitutability between labour and the other production inputs and, second, the degree of wage flexibility in the labour market. The latter effect leads to the compensation "via decrease in prices" discussed above.

Finally, the commercialisation of new ICT goods and services increases consumption and production and raises the demand for labour (compensation "via new products"). This effect would be larger the lower the substitutability of new products with existing ones and the higher the labour intensity of the production of the new products. In respect to the latter factor, one may expect the labour intensity of ICT products to decrease faster than in other industries, as ICT producing industries are the most intensive users of ICT process innovations.

This brief recollection of the predictions of the compensation theory suggests three main considerations. First, the impact of ICTs on employment is the result of opposing forces, which operate through a variety of channels, agents and industries. Looking only at some of these forces is likely to provide a biased assessment of the employment impact of ICTs.

Second, the mechanisms that are expected to compensate for the direct, negative effect of ICT process innovations on employment depend on several conditions that may not apply in reality, e.g. additional income generated by ICT process innovations may not be fully spent or invested, or that may take time to become effective, e.g. lower unit costs are not immediately translated into lower prices.

Finally, the compensation for the decrease in labour demand that may result from ICTs occurs through the mobility of resources—financial capital, knowledge assets and labour—across firms and sectors. By its very nature, this process of structural change takes time and may be hampered by institutional barriers and market imperfections. More fundamentally, entrepreneurial skills, intangible assets and workers' skills tend to be industry-specific and may not be fit to the business environment, the work organisation and the tasks composition of the activities where they would have to move. This is likely to be the case especially for new markets that did not exist before, like those created by new ICT goods and services.

2.3   Innovation and Employment: Findings from Recent Studies

Several empirical studies have analysed the relationship between innovation and employment. While only few of them focus on ICTs, their findings shade light on the effectiveness of the compensation mechanisms discussed in the previous section.

In the 1980s and 1990s, macroeconomic analysis dominated the research on the employment effects of innovation (e.g. Layard et al. 1994; Freeman and Soete 1994; Machin and Van Reenen 1998) whereas more recent analyses on this topic have been carried at the sectoral or firm level. Given the scope of the chapter, this section reviews the latest studies only (see Sabadash 2013 for a review of earlier studies).

In general, sectoral studies show that structural change is the driving force behind employment growth, with opportunities for both innovation and for jobs being sector-specific. Industry-level evidence for the 1990s and early 2000s in Europe suggests that the decrease in manufacturing employment was due to a combination of weak final demand, increasing wage and the prevalence of labour-saving process innovations over product innovations (Bogliacino and Pianta 2010; Bogliacino and Vivarelli 2011). Job losses occurred mostly in large firms, among low-skilled workers, in ICT and capital-intensive industries and in the financial sector. Job creation was concentrated in industries with high demand growth and those where product innovation dominated process innovation, as well as in open economies specialised in innovative and fast growing activities.

While the positive employment effects of product innovation are confirmed by firm-level studies, the effects of process innovations range from negative to positive according to the specification and the dataset.

A series of studies on European Community Innovation Survey (CIS) data based on a common micro-funded model (Peters 2004; Harrison et al. 2014) find out that employment losses are largely concentrated in non-innovating firms while employment growth is mainly driven by the introduction of new products. Process innovation was found to have negative employment effects only in German manufacturing industry.

Hall et al. (2008) run a similar model on a panel of Italian manufacturing firms over the period 1995–2003 and find positive employment effects for product innovation but no significant effect for process innovation.

Lachenmaier and Rottmann (2011) estimate a dynamic employment equation on a dataset of German manufacturing firms over the period

1982–2002. They find positive employment effects for different innovation measures, including process innovation.

Coad and Rao (2011) find out a positive correlation between employment and a composite innovativeness index (including both R&D and patents) in US high-tech manufacturing firms over the period 1963–2002.

Bogliacino et al. (2011) analyse a longitudinal database covering 677 European manufacturing and service firms over the period 1990–2008 and find a positive impact of R&D expenditures on employment in services and high-tech manufacturing but not in traditional manufacturing.

Finally, Evangelista and Vezzani (2012) find out that all types of innovation—including organisational innovation—affect employment indirectly by improving performances, leading to higher sales and more jobs. However, the classical distinction between product and process innovation is not able to capture these differentiated effects. Innovation strategies characterised by a combination of product, process and organisational innovations show the strongest positive impact on employment, whereas the negative direct effects of process innovations are found only in the manufacturing firms when process innovations are combined with organisational changes.

Different measures of innovation and ICTs are likely to explain to a large extent the different findings of these studies. In a study on Germany, Severgnini (2009) provides an interesting comparison among three different measures of ICTs: (1) a time trend, (2) the ratio of ICT investment to output, and (3) the contribution of ICTs to total factor productivity. These measures give opposite results. When ICTs are measured by a time trend, their employment effects tend to be negative in the short run and positive in the long run. However, long-run effects become statistically not significant when labour and product market regulations are controlled for. The second measure—the ratio of ICT investment to output—has mixed effects on employment while the third measure—the contribution of ICTs to total factor productivity—has negative effects in both the short and the long run.

While firm-level analyses permit a richer characterisation of innovation strategies and avoid the confounding effects from averaging different behaviours at the sectoral or macro-level, they miss out the employment effects that ICTs may have on other firms or industries.

First, firm-level databases are, in general, not representative of all firms and tend to be biased towards large manufacturing ones.

Second, micro-level studies do not distinguish whether employment growth in innovative firms results in net job creation—through "market expansion"—or it occurs at the expense of their rivals—through "business stealing." For instance, Greenan and Guellec (2000) show that the positive employment effects of process innovation found in French firms disappear at industry level.

Finally, when the business stealing effect is accounted for, firm-level analysis does not measure to what extent the same innovation that destroys jobs in one industry may result in job creation in a different industry via the compensation mechanisms discussed in Sect. 2.2.

Recent estimates of ICT employment multiplier based on input-output analysis suggest that these indirect effects are sizable. Such multipliers measure the overall increase in employment generated by 1 additional job in the ICT industry.

Katz (2012) reviews the broadband employment multiplier estimated by different studies: their value vary between 1.92 in Germany and 3.6 in the United States. Mandel and Scherer (2012) estimate that each new job in the mobile application industry generates another 0.5 jobs in the rest of the economy.

In their study of the employment impact of Facebook app development in the United States, Hann et al. (2011) use multipliers of 2.4 for the broadband industry, 2.5 for the communication sector and 3.4 for the whole economy.

Moretti (2012) argues that the high-tech job multiplier is as high as 5: for each job created in the software, technology and life-sciences industries in the United States, five new jobs are indirectly created in the local economy, 2 in high-skill occupations (e.g. doctors and lawyers) and 3 in low-skill occupations (e.g. waiters, barbers and store clerks).

Mazzolari and Ragusa (2013) find evidence of a strong positive relationship in the United States between the change in a city top-wage-bill share and the growth in local employment in jobs that substitute for home production. Consumption spillovers may account for one third of the growth of employment in home production substitutes experienced in the 1990s by non-college workers in the United States.

2.4 Modelling the Effects of ICT on Employment

This chapter analyses the effects of ICTs on employment within the standard labour demand theory (Hamermesh 1986). This framework has the advantage of modelling the employment effects of ICTs as a result of firms' decisions and market mechanisms rather than as a technology-driven outcome.

Fast technological progress in ICTs has led to a rapid decrease in the price of ICT equipment and software and to large investments in ICTs. Such investment have resulted into changes in the production mix of labour, ICT capital and other types of capital, on the one hand, and into a decrease in production costs and an increase in final demand, via lower prices and/or higher income, on the other.

The net impact of technological progress embodied in ICT capital on labour demand depends, therefore, on: (i) the extent to which ICT capital substitutes for labour (*partial elasticity of substitution*) and (ii) the extent to which lower unit costs generate higher demand and production via a decrease in prices (*price elasticity*) and/or an increase in income (*income elasticity*).

For the total economy, the economic theory predicts that both the price elasticity and the income elasticity of final demand are equal to one. Indeed, any decrease in the output price raises real income, thus leading to a proportional increase in real consumption and/or savings. Similarly, any increase in extra-profits raise nominal income, consumption and/or savings by the same proportion. By accounting identity, savings equals investments plus net exports. Therefore, any decrease in the output price and any increase in income would translate into an equal increase in final demand (consumption, investments and net exports).

It follows that ICT investments increase or decrease labour demand depending on whether the elasticity of substitution between labour and ICT capital is smaller or bigger than 1. The main aim of this study is, therefore, to estimate the value of the partial elasticity of substitution between labour and ICT capital. Spiezia (2018) provides a formal description of the model and its econometric specification.

This approach accounts for the employment effects of technological progress embodied in ICT capital goods but it does not consider disembodied technical change. The latter has effects both on the substitution between labour and ICT capital, on the one hand, and on the decrease in output price, on the other.

First, as discussed in the Sect. 2.2, disembodied technical change reduces the demand for labour per unit of output if it is labour-saving. Therefore, estimates based on embodied technical progress only may underrate the negative impact of ICT on employment. Second, disembodied technical progress raises multifactor productivity (MFP) thus reducing unit cost and output prices. Not accounting for disembodied technical change may, therefore, underestimate the positive effects of ICTs on final demand and employment.

While it is hard to quantify disembodied technical progress due to ICTs, two considerations suggest that the above measurement errors may not be large. First, there is growing evidence that: (i) a significant part of MFP is associated with investment in intangible assets (OECD 2013) and (ii) for ICT capital to raise productivity, it requires complementary investments in intangible assets (Corrado et al. 2014). Therefore, ICT investments are strongly correlated to intangible assets and are likely to capture a significant proportion of disembodied technical progress due to ICTs.

Second, firms' expectations about the future value of ICT capital services would also reflect productivity increases due to disembodied technical progress stemming from ICTs. As discussed in the following section, such expectations are reflected in ICT capital user costs and in the investment decisions by firms. Therefore, to the extent firms anticipated the productivity effects of disembodied technical progress, these effects would be also be captured by the estimates provided in this chapter.

2.5   The Dataset

The data for the analysis are drawn from the OECD Productivity Database (PDB), http://www.oecd.org/std/productivity-stats/. The PDB combines a consistent set of data on GDP, labour input, capital services, hourly wage and capital user costs for 19 countries over 1985–2012. The default source for the dataset is generally the OECD's Annual National Accounts, although other sources have been used when national accounts data were not available.

Labour input is defined as total hours worked of all persons engaged in production.

Capital inputs are measured as capital services: for any given type of asset, there is a flow of productive services from the cumulative stock of past investments. Capital service flows in the PDB relate to non-residential fixed capital only and can be broken down by seven types of assets: Hardware and office machinery; Communication equipment; Other machinery and equipment; Transport equipment; Non-residential construction; Software and Other products.

Estimates of capital services in the OECD PDB are based on the perpetual inventory method (PIM). The PIM calculations are carried out by the OECD, using an assumption of common service lives for given assets for all countries, and by correcting for differences in the national deflators used for hardware, communications equipment and software assets (Schreyer 2002;

Schreyer et al. 2003). The "harmonised" deflators assume that the ratios between ICT and non-ICT asset prices evolve in a similar manner across countries, using the United States as the benchmark.

The price of ICT capital services is the most important information for the purpose of this chapter. In general, the price of capital services is measured as their rental price. If there were complete markets for capital services, rental prices could be directly observed. This is, however, not the case for many capital goods that are owned and for which rental prices have to be imputed. The implicit rent that capital good owners "pay" themselves is defined as user costs of capital.

It is worth noticing that, unlike in other databases, e.g. EUKLEMS, the user cost of capital is not estimated by imposing the equality between capital remuneration and gross operating surplus (value added minus total wages) but it is based on firms' expectations about future capital productivity. Furthermore, this approach does not require perfect competition in the product market nor constant returns, e.g. to scale in production (Schreyer 2010).

Keeping aside more technical issues, two theoretical assumptions are crucial to the estimation of user costs. First, in a fully functioning asset market, the purchase price of an asset will equal the discounted flow of the value of services that the asset is expected to generate in the future (Jorgenson 1963). Second, a rational, cost-minimising producer will choose a vintage composition such that the relative productivity of different vintages is just equal to the relative user costs of the two vintages (Hulten 1990).

Changes in ICT user costs do not simply reflect improvements in technology but they also depend on firms' expectations about the future value of ICT capital services. Therefore, for a given ICT technological trend, country differences in the factors that affect these expectations, e.g. competition, regulation, cost of borrowing, consumer preferences, market size, etc. may affect the expected value of ICT capital services and the evolution of user costs.

Figure 2.2 shows the dynamics of the user cost of ICT capital over early 1990s–2012 for the three periods early 1990s–2001, 2001–2007 and 2008–2012. These periods correspond to three phases of the business cycle: before the dot.com bubble, after the subprime crisis and between the two crises.

Figure 2.2 shows two main trends. First, in all countries the decrease in ICT user costs has been faster in the second period (2001–2007) than in the first one (before 2001). The 2001–2007 decrease was the largest in Denmark, the Netherlands and Japan (about 10% a year). Second, in most countries, the decrease in ICT user costs has continued after the crisis but

Average yearly rate (%)

■ Before 2001   ■ Between crises   ■ After 2007

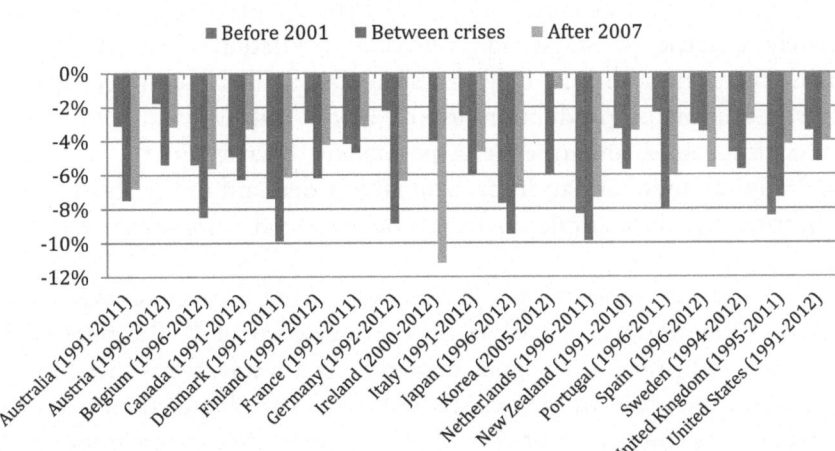

Fig. 2.2 Change in the user cost of ICT capital, 1990–2012 (*Source* Own calculations based on the OECD Productivity Database, 19 November 2015, http://www.oecd.org/std/productivity-stats/)

at a significantly slower rate. This slowdown is likely to reflect lower firms' expectations about future growth due to the crisis. Ireland and Spain are the only exceptions to this trend, as ICT user costs decreased at a faster rate than before.

2.6 Results

The partial elasticity of substitution between labour and ICT capital was estimated through system GMM (Blundell and Bond 1998) and detailed regression outputs are discussed in Spiezia (2018). The estimates provide two main results:

- In the long run, the estimated coefficient on ICT unit user cost is not statistically different from zero and the partial elasticity of substitution between labour and ICT capital is equal to one. Therefore, a permanent decrease in the user cost of ICT capital reduces labour demand per unit of output but it increases output by the same proportion. In other words, the substitution effect and the scale effect compensate each other completely. As a result, based on these estimates, investments in ICTs do not have any effect on labour demand in the long run.

- In the short run, however, firms cannot change production inputs immediately because of staggered contracts, regulations and other adjustment costs. In addition, ICT investments are likely to trigger a process of reallocation of production inputs across industries and this process takes time. As a consequence, a permanent decrease in the user cost of ICT capital does have an impact on labour demand in the short run. The adjustment path of employment can be described as follows.

In the first period, production techniques are fixed because it takes time for firms to change inputs. A decrease in the user costs of ICT capital leads to lower costs and prices and higher demand. As a result, firms hire more and employment increases. In the next period, firms can change their production technique. At a lower user of cost of ICT capital, they invest more in ICTs and reduce labour. As the hiring started in the first period is still producing its effects due to staggered contracts and adjustment costs, firms reduce employment below its long-run level. In the following periods, therefore, firms progressively increase employment as to bring it back to equilibrium.

The adjustment path following a permanent decrease in ICT user costs is illustrated in Fig. 2.3. The changes in employment are larger the larger the decrease in ICT user costs and the smaller the labour share in total costs. The return of employment to its long-run level is also slower the smaller the labour share. For the values of the labour share in OECD countries— between 0.65 and 0.88—the employment effects disappear after about 20 years.

Initial labour cost share is equal to the sample average in the first year (0.775)

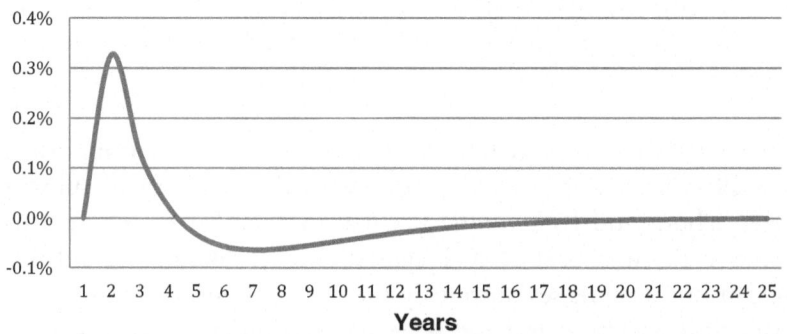

Fig. 2.3 Change (%) in labour demand following a permanent 5% decrease in the user cost of ICT capital (*Source* Own estimates based on regression output)

This dynamics is compounded by the fact that the permanent decrease in the user cost of ICT capital is not a "one-off" but it has been a continuous trend over more than two decades. Therefore, its employment effects have accumulated over time and become more persistent. In general, the employment effect of ICT remains positive for as long as the decrease in ICT user cost occurs at an increasing rate. When the decrease in ICT user cost slows down, the negative short-run effects of past capital accumulation prevail and result in a decrease in labour demand.

Figure 2.4 shows the change in employment driven by the accumulation of ICT capital over early 1990s–2012 for the three periods before 2001, 2001–2007 and 2008–2012, as discussed in Sect. 2.5.

The estimates suggest that ICT investments raised labour demand in all countries in both the period before 2001 and the subsequent period 2001–2007. In some countries, the cumulated contribution of ICT investments to employment growth over the two periods was significant: 7% in Denmark, Japan and the Netherlands; 6% in Germany; 5% in Australia, Belgium, Portugal and the United Kingdom; 4% in Canada, Finland, France, Italy, New Zealand, Sweden and the United States.

After 2007, ICTs have resulted in a decrease in labour demand in almost all countries. This seems due to the accumulation of short-run negative effects from past ICT investments and the slowdown in the decrease in ICT user costs and current ICT investments. The yearly decrease in labour demand after 2007, however, was much smaller than the yearly increase over

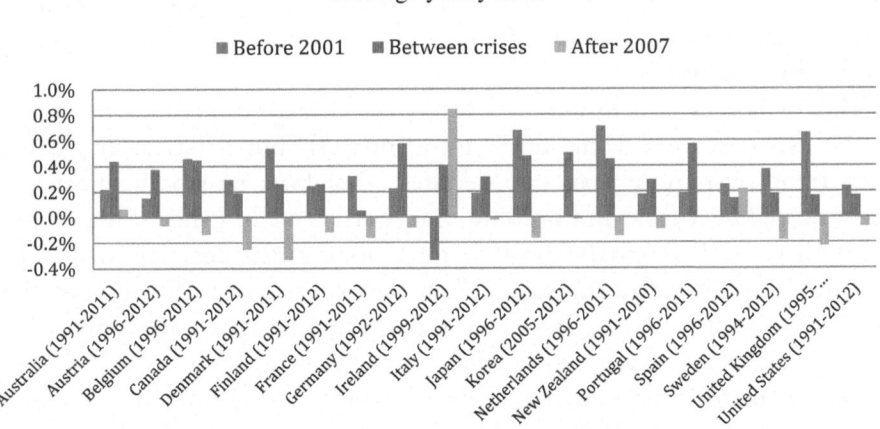

Fig. 2.4 Changes in labour demand due to growth in ICT capital (*Source* Own estimates based on regression output and the OECD Productivity Database, 19 November 2015, http://www.oecd.org/std/productivity-stats/)

the period early 1990s–2007. Therefore, the overall effect of ICT investment on labour demand over the whole period remains positive. Ireland, and to a lesser degree, Spain and Australia were the only countries where ICT investment led to an increase in labour demand after 2007.

As the post-2007 decrease in labour demand appears to be related to a slowdown in ICT investments, policies to foster such investments would be beneficial to employment. Also, the cost of temporary job losses due to the accumulation of past ICT investments could be relieved through labour market activation policies and temporary income support.

2.7   Conclusions

This study provides new estimates of the effect of ICT investments on labour demand in 19 OECD countries over the period early 1990s–2012. Its approach has been to measure ICT technical progress as the decline in the user cost of ICT capital and to estimate the effects of such decline on the demand for labour.

The findings suggest that ICT investments have no effects on labour demand in the long run. A permanent decrease in the user cost of ICT capital reduces labour demand per unit of output but it increases output by the same proportion. In other words, the substitution effect and the scale effect compensate each other completely.

In the short run, however, due to sluggish adjustments in production inputs, a one-off permanent decrease in ICT user cost results in a temporary increase in labour demand followed by a temporary decrease. Our estimates suggest that these temporary effects tend to disappear in about 20 years in most OECD countries.

This dynamics is compounded by the fact that the permanent decrease in the user cost of ICT capital is not a "one-off" but it has been a continuous trend over more than two decades. Therefore, its labour effects have accumulated over time and become more persistent. In general, the employment effect of ICT remains positive for as long as the decrease in ICT user cost occurs at an increasing rate. When the decrease in ICT user cost slows down, the negative short-run effects of past capital accumulation prevail and result in a decrease in labour demand.

Our estimates suggest that ICT investments raised labour demand in all countries in the early 1990s–2007 but reduced it afterwards. The decrease in labour demand after 2007, however, was smaller than the increase before

2007, thus leading to a positive effect of ICT investment on labour demand over the whole period.

While the negative employment effects of ICTs are estimated to fade away eventually, their persistence seem enough to justify appropriate policy measures, such as incentives to ICT investments, labour market activation policies and temporary income support.

This study has looked at the impact of ICT investments on the level of employment but not on its composition. Depending on data availability, the present framework could be extended to estimate labour demand for different types of skills or educational attainments.

Note

1. According to Hick's classification of technological progress.

References

Bogliacino, Francesco, and Mario Pianta. 2010. "Innovation and Employment: A Reinvestigation Using Revised Pavitt Classes." *Research Policy* 39 (6): 799–809. https://doi.org/10.1016/j.respol.2010.02.017.

Bogliacino, Francesco, and Marco Vivarelli. 2011. "The Job Creation Effect of R&D Expenditures." IZA Discussion Papers, No. 4728. IZA, Bonn, Germany. http://ftp.iza.org/dp4728.pdf.

Bogliacino, Francesco, Maria Cristina Piva, and Marco Vivarelli. 2011. "R&D and Employment: Some Evidence from European Microdata." IZA Discussion Papers, No. 5908. IZA, Bonn, Germany. http://ftp.iza.org/dp5908.pdf.

Blundell, Richard, and Stephen Bond. 1998. "Initial Conditions and Moment Restrictions in Dynamic Panel Data Models." *Journal of Econometrics* 87 (1): 115–143. https://doi.org/10.1016/S0304-4076(98)00009-8.

Coad, Alex, and Rekha Rao. 2011. "The Firm-Level Employment Effects of Innovations in High-Tech U.S. Manufacturing Industries." *Journal of Evolutionary Economics* 21 (2): 255–283. https://doi.org/10.1007/s00191-010-0209-x.

Corrado, Carol, Jonathan Haskel, and Cecilia Jona-Lasinio. 2014. "ICT, R&D and Non-R&D Intangible Capital: Complementary Relations and Industry Productivity Growth." Paper Presented at the 3rd World KLEMS Conference, May 19–20, 2014, in Tokyo. http://www.researchgate.net/publication/266209151.

Evangelista, Rinaldo, and Antonio Vezzani. 2012. "The Impact of Technological and Organizational Innovations on Employment in European Firms." *Industrial and Corporate Change* 21 (4): 871–899. https://doi.org/10.1093/icc/dtr069.

Freeman, Christopher, and Luc Soete. 1994. *Work for All or Mass Unemployment? Computerised Technical Change into the Twenty-First Century*. London and New York: Pinter.

Greenan, Nathalie, and Dominique Guellec. 2000. "Technological Innovation and Employment Reallocation." *Labour* 14 (4): 547–590. https://doi.org/ 10.1111/1467-9914.00146.

Hall, Bronwyn H., Francesca Lotti, and Jaques Mairesse. 2008. "Employment, Innovation, and Productivity: Evidence from Italian Microdata." *Industrial and Corporate Change* 17 (4): 813–839. https://doi.org/10.1093/icc/dtn022.

Hamermesh, Daniel S. 1986. "The Demand for Labour in the Long Run." In *Handbook of Labor Economics, Volume 1*, edited by Orley C. Ashenfelter and Richard Layard, 429–471. Amsterdam, the Netherlands: Elsevier. https://doi.org/10.1016/S1573-4463(86)01011-8.

Hann, Il-Horn, Siva Viswanathan, and Byungwan Koh. 2011. "The Facebook App Economy." Robert H. Smith School of Business, Center for Digital Innovation, Technology and Strategy, University of Maryland, September 19. https:// www.rhsmith.umd.edu/files/Documents/Centers/DIGITS/AppEconomy Impact091911.pdf.

Harrison, Rupert, Jordi Jaumandreu, Jacques Mairesse, and Bettina Peters. 2014. "Does Innovation Stimulate Employment? A Firm-Level Analysis Using Comparable Microdata from Four European Countries." *International Journal of Industrial Organization* 35 (July): 29–43. https://doi.org/10.1016/j. ijindorg.2014.06.001.

Hulten, Charles R. 1990. "The Measurement of Capital." In *Fifty Years of Economic Measurement*, edited by Ernst R. Berndt and Jack E. Triplett, 119–158. Chicago: University of Chicago Press. http://www.nber.org/chapters/c5974.pdf.

Jorgenson, Dale W. 1963. "Capital Theory and Investment Behaviour." *American Economic Review* 53: 247–259.

Katz, Raoul. 2012. "The Impact of Broadband on the Economy: Research to Date and Policy Issues." Broadband Series, ITU, Geneva, April. https://www.itu. int/ITU-D/treg/broadband/ITU-BB-Reports_Impact-of-Broadband-on-the-Economy.pdf.

Lachenmaier, Stefan, and Horst Rottmann. 2011. "Effects of Innovation on Employment: A Dynamic Panel Analysis." *International Journal of Industrial Organization* 29 (2): 210–220. https://doi.org/10.1016/j.ijindorg.2010.05.004.

Layard, Richard, Stephen Nickell, and Richard Jackman. 1994. *The Unemployment Crisis*. Oxford: Oxford University Press.

Machin, Stephen, and John Van Reenen. 1998. "Technology and Changes in Skill Structure: Evidence from Seven OECD Countries." *Quarterly Journal of Economics* 113 (4): 1215–1244. https://doi.org/10.1162/003355398555883.

Mandel, Michael, and Judith Scherer. 2012. "The Geography of the App Economy." South Mountain Economics, October. https://southmountaineco-nomics.files.wordpress.com/2012/11/the_geography_of_the_app_economy-f. pdf.

Mazzolari, Francesca, and Giuseppe Ragusa. 2013. "Spillovers from High-Skill Consumption to Low-Skill Labor Markets." *Review of Economics and Statistics* 95 (1): 74–86. https://doi.org/10.1162/REST_a_00234.

Moretti, Enrico. 2012. *The New Geography of Jobs*. Boston and New York: Mariner Books and Houghton Mifflin Harcourt.

OECD. 1994. *The OECD Jobs Study: Facts, Analysis and Strategy*. Paris: OECD.

OECD. 2013. *New Sources of Growth: Knowledge-Based Capital—Key Analyses and Policy Conclusions—Synthesis Report*. Paris: OECD.

Peters, Bettina. 2004. "Employment Effects of Different Innovation Activities: Microeconometric Evidence." ZEW Discussion Papers, No. 04-73. ZEW, Mannheim, Germany. ftp://ftp.zew.de/pub/zew-docs/dp/dp0473.pdf .

Sabadash, Anna. 2013. "ICT-Induced Technological Progress and Employment: A Happy Marriage or a Dangerous Liaison? A Literature Review." JRC Technical Reports, Institute for Prospective and Technological Studies, Digital Economy Working Paper, No. 76143. IPTS, Joint Research Centre, Seville, Spain. https://doi.org/10.2791/2141.

Schreyer, Paul. 2002. "Computer Price Indices and International Growth Comparisons." *Review of Income and Wealth* 48 (1): 15–31. http://dx.doi.org/10.1111/1475-4991.00038.

Schreyer, Paul. 2010. "Measuring Multi-Factor Productivity When Rates of Return Are exogenous." In *Price and Productivity Measurement: Volume 6—Index Number Theory*, edited by W. Erwin Diewert, Bert M. Balk, Kevin J. Fox, Dennis Fixler, and Alice O. Nakamura, 13–40. Bloomington, IN: Trafford Publishing.

Schreyer, Paul, Pierre-Emmanuel Bignon, and Julien Dupont. 2003. "OECD Capital Services Estimates: Methodology and a First Set of Results." OECD Statistics Working Papers, No. 6. OECD Publishing, Paris. http://dx.doi.org/10.1787/658687860232.

Severgnini, Battista. 2009. "Growth Accounting, ICT and Labour Demand." In *The Impact of ICT on Employment*, edited by Michael Burda, 1–61. Berlin, Germany: Humbold-Universitat zu Berlin. https://publications.europa.eu/en/publication-detail/-/publication/2e16f508-1acc-4042-afd1-19ce3c78e841.

Spiezia, Vincenzo. 2018. "ICTs and Jobs: Complements or Substitutes? The Effects of ICT Investment on Labour Demand in 19 OECD Countries." *OECD Journal: Economic Studies* (Forthcoming).

Spiezia, Vincenzo, and Marco Vivarelli. 2002. "What Do We Know About the Effects of Information and Communication Technologies on Employment Levels?" In *Productivity, Inequality, and the Digital Economy—A Transatlantic Perspective*, edited by Nathalie Greenan, Yannick L'Horty, and Jaques Mairesse. Cambridge, MA: MIT Press.

3

A One-Sector Model of Robotic Immiserization

Jeffrey D. Sachs, Seth G. Benzell and Guillermo Lagarda

3.1 Introduction

The word robot comes from the Czech word "robota", meaning forced labor. Ever since the term's invention by Karl Čapek in his 1920 dystopian science fiction masterpiece R.U.R, it has been associated with ambivalence about the power of automation. The play begins with the general manager of Rossum's Universal Robots discussing the potential of his assembled beings to raise living standards. He predicts that his robot laborers will lower the prices of goods to zero, ending toil and poverty forever. This plan hits a small snag when the robots decide to overthrow their masters and destroy all humans. But was the manager's economic forecast even correct in the first place?

This paper investigates the implications of capital investments, in the form of robots, which allow for production without labor. Our key finding is that an increase in robotic productivity will temporarily raise output, but, by lowering the demand for labor, can lower wages and consumption in

J. D. Sachs
Center for Sustainable Development, Columbia University,
New York, NY, USA

S. G. Benzell
MIT Initiative on the Digital Economy, Boston University, Boston, MA, USA

G. Lagarda (✉)
Inter-American Development Bank and Global Development Policy Center, Boston University, Boston, MA, USA

© The Author(s) 2018
L. Pupillo et al. (eds.), *Digitized Labor*, https://doi.org/10.1007/978-3-319-78420-5_3

the long run. In what we term a *paradox of robotic productivity*, innovations that increase the productivity of robotic investments can, after a generation, lower robotic and total output, and lower the well-being (lifetime utility) of all future generations. The mechanism for this immiserization is decreased wages of the workers with whom the robots compete. We find this immiserization is most likely when the future is heavily discounted, goods produced by robots are close substitutes for goods created by human labor, and when traditional capital is a more important factor in non-robotic production (so that the reduction of traditional capital has a larger adverse impact on wages). In our richest setting, increases in robotic productivity lower well-being until a threshold is reached. After reaching the threshold, the economy may grow indefinitely.

The fact that a rise in robotic productivity can immiserize future generations may seem paradoxical. After all, higher productivity enables society to produce more output from the same quantity of inputs. If the market response to robotic innovations does not lead to a positive result, this suggests that there may be a role for government intervention. We show this intuition to be correct. Immiserization may be overcome through redistributive policies of the state.

3.2   Literature Review

Even before the birth of modern science fiction, academics and ordinary people have been concerned about the potential downsides of technological growth.[1] The English Luddites of the late eighteenth and early nineteenth centuries famously organized raids and riots against the industrial machines they felt were taking their jobs. In the second half of the nineteenth century, Marx (1992) bemoaned the fact that under capitalism "all methods for raising the social productivity of labor are put into effect at the cost of the individual worker". In the first half of the twentieth century, Keynes (1933) cautioned against overreaction to "technological unemployment", which, while painful for displaced workers, was merely a "temporary phase of maladjustment". Similarly, Schumpeter (1939) championed the "creative destruction" of capitalism, in which older ways of doing work are, not without pain, superseded by advances in technology as new types of more productive work are created.

In the economic prosperity of the postwar era, the views of technological optimists generally held sway. However, recent wage stagnation and growing inequality across the developed world have led economists to take another hard look at technological growth. Autor et al. (2003) Acemoglu and Autor (2011), and Autor and Dorn (2013) trace recent declines in employment and wages of middle-skilled workers to the development of smart machines. Katz and Margo (2014) point to similar *labor polarization* during the early stages of America's industrial revolution. Goos et al. (2010) offer additional supporting evidence for Europe. Sachs and Kotlikoff (2012) present a model in which robots immiserize future generations, a precursor of the models studied in this paper. However, Mishel et al. (2013) argue that "robots" can't be "blamed" for post-1970s US job polarization given the observed timing of changes in relative wages and employment. A literature inspired by Nelson and Phelps (1966) hypothesizes that inequality may be driven by skilled workers more easily adapting to technological change, but generally predicts only transitory increases in inequality.

A potential implication of our model is a decline, over time, in labor's share of national income. US national accounts record a stable percent share of national income going to labor during the 1980s and 1990s. But starting in the 2000s labor's share has dropped significantly. Frey and Osborne (2013) try to quantify prospective human redundancy arguing that over 47% of current jobs will likely be automated in the next two decades. Olsen and Hemous (2014) calibrate a model in which capital can substitute for low-skilled labor while complementing high-skilled labor to explain trends in the labor share of income and inequality.

The lessons of our model are also related to the endogenous growth literature. In Rebelo's (1991) AK model, sustained per capita output growth occurs so long as there are no decreasing returns to scale in production. This model complemented Romer (1990) which included open-ended growth driven by endogenous technological development in the tradition of learning by doing proposed by Arrow (1962).

There are several models that include a potential for welfare-improving intergenerational transfers. Two papers that with mechanisms more similar to this one are Sachs and Kotlikoff (2012) and Benzell et al. (2015). These papers also posit that technological changes may immiserize future generations through the mechanism of reduced wages.

3.3   The Model Framework

The essential quality of robots, as we define them, is that they allow for output without labor. To produce a unit of output from robotic technology, entrepreneurs need only make a capital investment. Innovation in robotic production can therefore change labor's share of national income. In a model with an infinitely lived representative consumer, this is unlikely to have major effects. However, if those earning labor and capital income have different propensities to consume, then a change in labor's share of income can have important effects on saving and investment. We attempt to capture this effect in the simplest possible setting.

The setup is an overlapping generations (OLG) model with two cohorts. This allows for labor's share of income to have a dynamic effect and straightforward generational welfare analysis.

3.3.1 Households

All individuals live for two periods, working, saving and consuming while young, and consuming while old. Workers in this economy maximize a lifetime utility function of the form

$$U_t = \phi u(\vec{c}_{1,t}) + (1 - \phi)u(\vec{c}_{2,t+1}), \tag{3.1}$$

where $\vec{c}_{1,t}$ and $\vec{c}_{2,t+1}$ are vectors of goods consumed by a household in the first and second periods of life, and $u(\cdot)$ is a within-period homothetic utility function. Henceforth, we assume within-period utility is logarithmic, $u(\vec{c}_t) = \ln(v(\vec{c}_t))$, where v is Cobb–Douglas with constant returns to scale. There is no leisure.

A generation maximizes U_t subject to its lifetime budget constraint, which in general may include government taxes and transfers.

$$w_t L_t + G_t = \vec{p}_t \vec{c}_{1,t} + \frac{\vec{p}_{t+1} \vec{c}_{2,t+1}}{1 + [r_{t+1}(1 - \tau_t)]}, \tag{3.2}$$

where $\vec{p}_t$ is a vector of prices, w_t is the wage, G_t is the size of government grants to the young, $1 - r_t$ the interest rate, and τ_t is the capital income tax rate. For convenience, define the net income of the young as the sum of their labor income and any government transfer, and the net interest rate of the old as net of the government capital income tax. Thus,

$$w_t^N = w_t L_t + G_t, \tag{3.3}$$

and

$$r_t^N = r_t(1 - \tau_t). \tag{3.4}$$

Utility maximization leads to the well-known result that saving, S_t, equals a fixed fraction $(1 - \phi)$ of youth income,

$$S_t = (1 - \phi)(w_t^N). \tag{3.5}$$

Households allocate savings with perfect foresight between available types of physical assets to maximize returns.

3.4 Production

There are two perfectly competitive types of firms. Time t production of the consumption and investment good with the traditional output technology, $X_{m,t}$, is as follows

$$X_{m,t} = D_{X,t} M_{X,t}^\epsilon L_{X,t}^{1-\epsilon}, \tag{3.6}$$

where $M_{X,t}$ is the amount of machines rented by these firms, $L_{X,t}$ is the amount of labor hired, ϵ is a Cobb–Douglas parameter, and $D_{X,t}$ a total factor productivity term. Production by robotic firms is as follows

$$X_{r,t} = \theta_t R_t, \tag{3.7}$$

where $X_{r,t}$ is the output of these firms, R_t is the amount of robots rented by these firms, and θ_t is the robotic productivity. Factor demands for robots, machines, and labor reflect profit maximization

$$\max{}_{M_{X,t} L_{X,t}} X_{m,t}(M_{X,t}, L_{X,t}) - w_t L_{X,t} - m_t M_{X,t} \tag{3.8}$$

and

$$\max{}_{R_t} X_{r,t}(R_t) - \rho_t R_t, \tag{3.9}$$

where m_t is the rental rate for machines and ρ_t is the rental rate for robots.
These yield the first order conditions

$$w_t = (1 - \epsilon) D_{X,t} M_{X,t}^\epsilon L_{X,t}^{-\epsilon}, \tag{3.10}$$

$$m_t = \epsilon D_{X,t} M_{X,t}^{\epsilon-1} L_{X,t}^{1-\epsilon}, \tag{3.11}$$

and

$$\rho_t = \theta_t. \tag{3.12}$$

3.4.1 Households

Utility is logarithmic in consumption of the one good.

$$u(x_t) = \ln(x_t),$$
(3.13)

Household demands for consumption and investment satisfy

$$x_{1,t} = \phi w_t^N$$
(3.14)

and

$$x_{2,t} = (1 + r_t^N)K_t,$$
(3.15)

where K_t is capital of any type owned by the old.

3.4.2 Equilibrium

The total output of the economy is the sum of the outputs of the two types of firms,

$$X_t = X_{m,t} + X_{r,t}.$$
(3.16)

The one-sector model is in equilibrium when the market for goods clears,

$$X_t = x_{1,t} + x_{2,t} + S_t,$$
(3.17)

the labor market clears,

$$L_{X,t} = L_t,$$
(3.18)

the government is balancing its budget,

$$G_t = r_t \tau_t K_t,$$
(3.19)

and the market for investments clears,

$$S_t = K_{t+1} = M_{X,t+1} + R_{X,t+1},$$
(3.20)

as capital depreciates fully each period.

Finally, investment seeks maximum returns in the subsequent period with perfect foresight. Here we are only interested in the case where robots are productive enough to be used, so investment must equalize the rate of return of both forms of capital. Therefore,

$$1 + r_t = m_t = \rho_t = \theta_t.$$
(3.21)

3.4.3 Equilibrium Analysis

Consider the case where $D_{X,t} = 1$ and $L_t = 1$ in all periods.

Combining first order equations yields

$$w_t = (1 - \epsilon)\frac{e^{\frac{1}{1-\epsilon}}}{\theta_t} \qquad (3.22)$$

Note that a rise in robot productivity reduces the wage. The reason is that higher θ shifts investment from machines into robots. This lowers the capital-labor ratio in X_m firms, decreasing the marginal productivity of workers. The wage is not influenced by the capital stock, because both the quantity of labor and the interest rate are fixed by factors outside the traditional firms. This in turn fixes the amount of capital in traditional firms and therefore the wage.

We can write the indirect utility function in terms of θ_t and θ_{t+1}. Ignoring constant terms, and assuming no transfers $(G_t = \tau_t = 0)$ we have

$$U_t = \ln w_t + (1 - \phi)\ln(1 + r_{t+1}), \qquad (3.23)$$

or equivalently,

$$U_t = \frac{-\epsilon}{1 - \epsilon}\ln\theta_t + (1 - \phi)\ln\theta_{t+1}. \qquad (3.24)$$

Notice that robot productivity has two opposing effects on lifetime utility. High θ_t lowers the wage while high θ_{t+1} raises the returns to saving. The negative wage effect tends to dominate the saving effect when the capital share of income (ϵ) in traditional firms is large, because this measures the importance of machines in complementing the labor or workers. Immiserization is also more likely when the discount rate ϕ is higher, because a high ϕ means that the utility value of higher returns to saving is low.

Consider a one-step permanent rise of θ at time T. That is for $t < T$, $\theta_t = \theta^L$ and for $t \geq T, \theta_t = \theta^H > \theta^L$. The lifetime utility of an individual born in t is

for $t < T - 1$

$$U_t = \frac{-\epsilon}{1 - \epsilon}\ln\theta^L + (1 - \phi)\ln\theta^L, \qquad (3.25)$$

when $t = T - 1$

$$U_t = \frac{-\epsilon}{1-\epsilon}\ln\theta^L + (1-\phi)\ln\theta^H, \tag{3.26}$$

and if $t > T-1$

$$U_t = \frac{-\epsilon}{1-\epsilon}\ln\theta^H + (1-\phi)\ln\theta^H. \tag{3.27}$$

The rise in robot productivity in period T must raise the welfare of generation $T-1$. For that generation, the rise of robot productivity was too late to impact their wage. However, the return on their saving is increased by the rise in robotic productivity in period T. Generation $T-1$, in other words, will enjoy high wages when young and high retirement income when old. Generations T and after will not be so lucky. For them, the positive effect of better robots is at least partially offset by lower wages.

An increase in robotic productivity will induce long-run immiserization[2] as long as

$$\frac{\epsilon}{1-\epsilon} > (1-\phi). \tag{3.28}$$

If Eq. 3.28 holds, the wage effect dominates and leads to a decline in lifetime utility. Only a single generation benefits from the rise of robot productivity, specifically the generation born in the period before the improvement in robot productivity. That generation benefits from higher returns to saving without incurring the negative shock of lower wages.

3.4.4 Ensuring That All Generations Benefit from the Rise in θ

Could a managed rise of robots lead to a better long-run outcome? It is clear that markets alone are not sufficient to ensure that a rise of robot productivity raises the well-being of future generations. However, it seems likely that a pure rise in productivity, by pushing out the production possibility frontier, can be made into a rise in lifetime utility for all generations with the right kind of government intervention. To insure a better outcome, the income of the young should be augmented by redistribution from the old.

Here's how to turn the robotics innovation in time T into a rise in well-being for all generations from time $T-1$ onward.

In every period T and after, the government levies a tax on the capital income of retirees and transfers the proceeds as a grant G_t to the young.

Let the government set the grant equal to the decline of the wage caused by the rise of θ. Let w^H be the market wage associated with θ^H and w^L be the market wage associated with θ^L. Then necessarily, $w^L > w^H$. The grant mechanism will function as follows: For $t > T - 1$

$$G_t = w_t^L - w_t^H. \tag{3.29}$$

To pay for this grant, the government levies a capital-income tax at rate τ_t on the old in each period. With saving S_t, pre-tax capital income is given by $\theta^H S_t$. Therefore, the tax rate should be set such that for $t \geq T$

$$G_t = \left(\theta^H - 1\right)\tau_t K_t. \tag{3.30}$$

Of course, savers anticipate this capital income tax and plan their inter-temporal spending decisions accordingly. Instead of earning a rate of return θ^H, savers will earn a net-of-tax rate of return $\left(1 + r_{t+1}^N\right) = 1 + (\theta^H - 1)(1 - \tau_t)$. Because of their logarithmic preferences this change in rate of return does not change their saving behavior. The indirect lifetime utility function can be rewritten in terms of youth net-of-transfer income w_t^N and r_{t+1}^N. Since policy fixes the disposable wage at w_t^L we have, ignoring constant terms,

$$U_t^L = \ln(w_t^L) + (1 - \phi)\ln(1 + r_{t+1}^N). \tag{3.31}$$

Every generation will be better off when θ rises to θ^H, as net-of-tax lifetime budget constraints must be larger than when θ^L.

When θ rises, it is easy to see that X_t rises instantaneously as well. This is because the level of capital is unchanged, but its productivity has increased. Now, consider total output from the perspective of factor income. Since there are no profits, $X_{r,t} = \theta R_t$ and $X_{m,t} = w_t + \theta M_t$, we have that $X_t = w_t + \theta(R_t + M_t) = w_t + \theta S_{t-1}$. By Eq. 3.5, S_t depends only on the net income of the young w_t^N. The transfer system keeps the disposable wage equal to w_t^L, so saving S_t also remains unchanged when θ rises. When θ rises, the overall rise of X_t ensures that $w_t^H + \theta^H S_t > w_t^L + \theta^L S_t$. Therefore, $w_t^H - w_t^L + \theta^H S_t > \theta^L S_t$. Since $w_t^L - w_t^H$ equals G_t, which is also equal to $(1 + (\theta^H - 1)\tau_t)S_{t-1}$, we find that $(1 + (\theta^H - 1)\tau_t)S_{t-1} > \theta^L S_t$. Hence, $(1 + r_{t+1}^N) = (1 + (\theta^H - 1)\tau_t) > \theta^L$.

This reasoning establishes a key result. By taxing the capital of the old, and transferring the proceeds to the young, the government keeps the net income of the young unchanged while the net-of-tax rate of return on saving is higher. Therefore, the rise of robot productivity to θ^H combined with the fiscal transfer system raises the well-being of all generations compared with the utility when productivity equals θ^L.

The result is important in light of discussions as to whether robotics will necessarily raise or lower well-being. The answer is that higher productivity is a potential gain for all generations, but only if government undertakes redistributive policies to ensure that indeed all generations benefit. Without such redistribution, it is possible, we have seen, that the robotics innovation improves the well-being of just one generation, while lowering the lifetime well-being of all future generations.

3.5   Conclusion

The rise of the robots is already creating major disruption in labor markets, essentially turning production processes more capital intensive. When robots are close substitutes for production by labor and machinery, the demand for labor is likely to decline, threatening a decline of wages, saving, and economic well-being of current and future generations. We have qualified that intuition, however. Government redistribution can ensure that a pure productivity improvement raises well-being of all generations. In the example shown in the paper, government taxes the capital owned by retirees and distributing the proceeds to young workers.

Notes

1. This section draws on Benzell et al. (2015).
2. On the other hand, a reduction in long-run national consumption can only occur if θ increases above 1. This is because the golden rule (long-run consumption maximizing) level of saving, given constant L and 100 percent depreciation is that which brings long-run interest rates equal to 1. In cases where θ increases from a level below 1 to a level closer to but still below 1, long-run consumption will increase although welfare may decrease.

References

Acemoglu, Daron, and David Autor. 2011. "Skills, Tasks and Technologies: Implications for Employment and Earnings." *Handbook of Labor Economics* 4: 1043–1171.

Arrow, K. 1962. "The Economic Implications of Learning by Doing." *The Review of Economic Studies* 29 (3): 155.

Autor, David H., and David Dorn. 2013. "The Growth of Low-Skill Service Jobs and the Polarization of the US Labor Market." *The American Economic Review* 103 (5): 1553–1597.

Autor, David H., Frank Levy, and Richard J. Murnane. 2003. "The Skill Content of Recent Technological Change: An Empirical Exploration." *The Quarterly Journal of Economics* 118 (4): 1279–1333.

Benzell, Seth G., Laurence J. Kotlikoff, Guillermo LaGarda, and Jeffrey D. Sachs. 2015. *Robots Are Us: Some Economics of Human Replacement*. No. w20941. National Bureau of Economic Research.

Frey, Carl Benedikt, and Michael A. Osborne. 2013. *"The Future of Employment."* Oxford Martin.

Goos, Maarten, Alan Manning, and Anna Salomons. 2010. *Explaining Job Polarization in Europe: The Roles of Technology, Globalization and Institutions*. No. dp1026. Centre for Economic Performance, LSE.

Katz, Lawrence F., and Robert A. Margo. 2014. "Technical Change and the Relative Demand for Skilled Labor: The United States in Historical Perspective." In *Human Capital in History: The American Record*, pp. 15–57. Chicago: University of Chicago Press.

Keynes, John Maynard. 1933. "Economic Possibilities for Our Grandchildren (1930)." In *Essays in Persuasion*, pp. 358–373. New York: Harcourt Brace.

Long, Tony. 2017. "Jan. 25, 1921: Robots First Czech In." *WIRED* [online]. Available at: http://www.wired.com/2011/01/0125robot-cometh-capek-rur-debut/.

Marx, Karl, and Ben Fowkes. 1992. "A Critique of Political Economy. Vol. 1 of Capital."

Mishel, Lawrence, John Schmitt, and Heidi Shierholz. 2013. "Assessing the Job Polarization Explanation of Growing Wage Inequality." Working Paper, Economic Policy Institute.

Nationalarchives.gov.uk. 2017. The National Archives Learning Curve | Power, Politics and Protest | Luddites [online]. Available at: http://www.nationalarchives.gov.uk/education/politics/g3/.

Nelson, Richard R., and Edmund S. Phelps. 1966. "Investment in Humans, Technological Diffusion, and Economic Growth." *The American Economic Review* 56 (1/2): 69–75.

Olsen, Morten, and David Hemous. 2014. "The Rise of the Machines: Automation, Horizontal Innovation and Income Inequality." In *2014 Meeting Papers*, no. 162. Society for Economic Dynamics.

Rebelo, Sergio. 1991. "Long-Run Policy Analysis and Long-Run Growth." *Journal of Political Economy* 99 (3): 500–521.

Romer, Paul M. 1990. "Endogenous Technological Change." *Journal of Political Economy* 98 (5, Part 2): S71–S102.

Sachs, Jeffrey D., and Laurence J. Kotlikoff. 2012. *Smart Machines and Long-Term Misery*. No. w18629. National Bureau of Economic Research.
Schumpeter, Joseph Alois. 1939. *Business Cycles*, vol. 1. New York: McGraw-Hill.

4

Routinization and the Labour Market: Evidence from European Countries

Federico Biagi, Paolo Naticchioni, Giuseppe Ragusa
and Claudia Vittori

4.1 Introduction

The impact of technological change on the labour market is an evergreen topic in economics, due to an endless list of contributions started with David Ricardo, Karl Marx, John Maynard Keynes, and Wallise Leontief, among others. In this chapter, we take into account a recent debate about a specific dimension of the technological change, concerning the routinization process that is taking place in most of the developed and developing countries, and its impact on the labour markets of European countries.

This interesting issue has been introduced at the core of the recent economic and policy debate by some recent best seller books, such as Brynjolfsson and McAfee (2011, 2014), which have provided some forecasts on how technological progress will impact ordinary life of individuals, in particular on the labour market. According to Brynjolfsson and McAfee (2011, 2014),

F. Biagi
University of Padova, Padova, Italy

P. Naticchioni (✉)
Roma Tre University, Rome, Italy

G. Ragusa
Luiss Business School, Rome, Italy

C. Vittori
Sapienza University of Rome, Rome, Italy

© The Author(s) 2018
L. Pupillo et al. (eds.), *Digitized Labor*, https://doi.org/10.1007/978-3-319-78420-5_4

the technological progress is entering into a new historical phase that will completely change the existing standards for individuals and firms. Also, Brynjolfsson and McAfee (2011, 2014) stress the importance of the routinization trends as a new shape of the technological change, and its impact on the labour market.

Even if this is a recent phenomenon, there is already a robust literature on routinization, which has often been associated to the polarization trends in the labour market.

Autor et al. (2006) have shown that the US labor market has become increasingly polarized both in terms of occupations and wage distributions. The routinization hypothesis has been advocated for explaining this empirical evidence (Autor et al. 2003). Goos et al. (2009) show similar findings for Europe, where the fall in the middling occupations, the ones mostly routinized, has driven the job polarization processes.

So far, most of the literature has investigated these issues making comparisons between cross-sectional data over time, analysing changes in employment share in each occupation (manual/service, routine, and abstract), without analysing the flows, i.e. individual longitudinal careers. Cortes (2016) and Cortes et al. (2016) analyse job polarization using a dynamic approach for the United States, highlighting the importance of using longitudinal data: Flows are crucial to understand the mechanisms behind routinization and job polarization.

In this chapter, we focus on the routinization processes in Europe, using the Survey on Income and Living Conditions (SILC) data. The main reason for using these data is their longitudinal dimension, which allows us to follow individual careers over time. Using longitudinal data, we focus on the following issues. First, we investigate whether routinization processes are confirmed. Second, we study the determinants of routinization, using the rich EU-SILC set of available socio-economic variables. Third, we analyse the relation between routinization and unemployment inflows, to investigate whether routinization can be considered as a driver of unemployment.

We consider several routinization variables, in line with the literature. In such a way, we aim at testing whether using different measures may lead to different outcomes, from both a qualitative and quantitative point of view. Furthermore, apart from studying the sample with all available European countries, we also focus on five different groups of countries (Southern, Continental, Anglo-Saxon, Nordic, and Eastern), which are characterized by different institutions, cultural settings, business cycles, and so on, in order to detect any heterogeneity in the routinization process.

4.2   The EU-SILC Database

We make use of European data, the SILC, from 2004 to 2012. We do not use data for the following years (2013 and 2014) since the classification of occupation ISCO has substantially changed in the SILC data in 2012, from the ISCO88 classification to the ISCO08. This means we can use the ISCO88 classification until 2011, while we use the year 2012 only to compute some employment status that will be used in the analysis.

The SILC data have been introduced in 2004, in order to replace the European Community Household Panel that has been in place from 1994 to 2001. The SILC data are very rich in terms of information. The primary focus of SILC is the collection of information on the income and living conditions of different types of households in order to derive indicators on poverty, deprivation, and social exclusion. It is a voluntary survey (for potential respondents) of private households and is carried out under EU legislation (Council Regulation No. 1177/2003). The income reference period for SILC is the 12 months prior to date of interview. SILC is currently being conducted on an annual basis in order to monitor changes in income and living conditions over time. The target population is the representative sample of households in the different countries, and the sample size is different across countries.

The EU-SILC panel is a rotational panel which is comparable in its structure to the Current Population Survey. In a rotational panel, the same individuals are interviewed for a certain time period (four years) and each year one-quarter of all respondents are replaced by new respondents. The integrated design consists in selecting four panels at the first wave. Each subsequent year, a panel is dropped and replaced by a new replication. For a given year, the respective longitudinal file available from Eurostat only contains those respondents that were interviewed both in the respective year and in the preceding years. In this work, we follow Engel and Schaffner (2012), merging different panel data sets provided by Eurostat. Hence, we derive a panel over the whole period 2004–2012.

In addition to the structure of the data, the cross-sectional data and the longitudinal data also differ to some extent in the covered variables. There are some variables in the cross-sectional data file that are also of interest for the analysis of labour market transitions and mobility, but they are not included in the longitudinal data sets. For our analysis, the main limitation of using the longitudinal data is related to the fact that while the occupation variable is available, the industry classification variable is missing.

This is another disappointing choice made by Eurostat, that really represents a strong constraint in any analysis of the labour market. In addition, we do not consider the sample for Germany, since there is evidence that German EU-SILC data on income is plagued with problems (Hauser 2008; Frick and Krell 2010).

It is interesting to note that so far the literature on the impact of technological change on the labour market has mainly focused on comparisons between cross-sectional data over time, investigating changes in employment share in each occupation (manual/service, routine, and abstract), without analysing the flows, i.e. individual longitudinal careers. Cortes (2016) and Cortes et al. (2016) analyse job polarization using a dynamic approach for the United States, highlighting the importance of using longitudinal data: Flows are crucial to understand the mechanisms behind job polarization.

Cortes (2016) shows that workers in routine jobs have higher probability to separate from their job. Interestingly, he finds that: High skilled routine workers display higher probability to move to abstract jobs; low ability routine workers show higher probability to switch to manual jobs; the decrease in routines employment is primarily due to the change in propensity of young workers to enter the labour market in routine jobs.

Furthermore, in order to investigate the potential heterogeneity across European countries, we carry out our analysis at the European level as a whole, considering all countries in the data sample, as well as at the level of standard groups of countries for Europe, defined as follows: Southern (Italy, Spain, Portugal, Greece, and Cyprus); Continental (Austria, Belgium, France, Luxembourg, and the Netherlands); Anglo-Saxon (Great Britain, Ireland); Nordic (Denmark, Finland, Island, Norway, and Sweden); and Eastern (Bulgaria, Check Republic, Poland, Romania, Slovakia, Lithuania, Latvia, Slovenia, and Estonia).

4.2.1 Different Measures of Routinization

In the literature, several measures of routinization are available, but none of the measures can be so far considered as the standard, the best one to use. For this reason, in this work, we will make use of different measures of routinization, in order to provide a test of consistency across different measures. The first variable is the Routine Task Intensity (RTI) index, developed by Autor and Dorn (2013). At first, they measure routine task activities using the occupational composition of employment. Following Autor et al. (2003), they merge job task requirements from the fourth edition of the US Department of Labor's Dictionary of Occupational Titles (DOT) to their corresponding Census occupation classifications to measure routine, abstract, and manual task content by occupation. These three variables measure intensity in the different type of task. They hence combine these measures to create a summary measure of RTI that varies at occupation classification ISCO 2-digits, standardized in the whole data set (zero mean and one standard deviation), as follows:

$$RTI = \ln(\text{Routine}) - \ln(\text{Manual}) - \ln(\text{Abstract})$$

This measure increases in the importance of routine tasks in each occupation and decreases in the importance of manual and abstract tasks.

As second measures, we use directly the three intensity measures in abstract, routine, and manual/service tasks, that have been aggregated to compute the RTI index, i.e. the task measures used in Autor et al. (2003).

4.2.2 Trends in Routinization in Europe

In the following we present routinization trends in Europe, and in the different groups of countries. The period of analysis provided by the EU-SILC data is 2004–2011. As mentioned earlier, we cannot consider years after 2011 since the occupation classification has completely changed from ISCO88 to ISCO08.

In Fig. 4.1, we show the trend over time of the RTI measure that will be the main variable used in the paper because of its synthetic nature. RTI displays a clear decreasing trend overtime, suggesting that the relative intensity

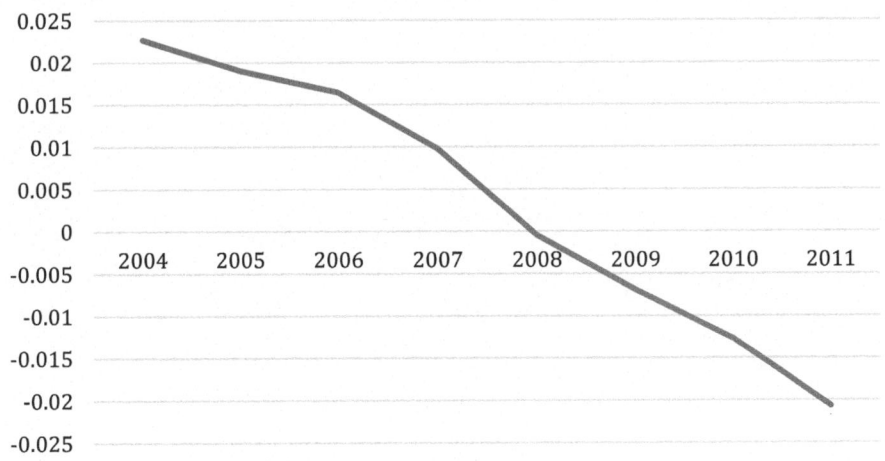

Fig. 4.1 Trends of the RTI index. All countries

of routine jobs are decreasing in the labour market, consistently with the intuition that routine jobs can be more easily replaced by technology. The magnitude of the change is relevant but not huge when considering that the RTI by definition is constrained to have zero mean and one standard deviation.

Figure 4.2 instead includes the trends of the three separate task measures described above: Abstract, routine, and manual (used to compute the RTI index). Actually using these measures, the routinization trends are even more pronounced. As expected, the abstract intensity increases sharply overtime, the routine one decreases, and the manual intensity has a non-monotonic trend, increasing until 2008 and then decreasing. Hence, only a part of the routinization story applies, since routine intensity decreases but this is not associated to a monotonic increase in manual intensity tasks.

Let us now move to the evidence concerning different set of countries. In Fig. 4.3, we include the trends of the RTI intensity in the five set of countries defined above. First of all, it is worth noting that the RTI variable is standardized in the sample of all European countries in all years. This means that differences across lines in Fig. 4.4 indicate differences in intensity levels across groups. In such a framework, it emerges that trends for all groups display a similar decreasing trend across groups. However, the Nordic group line is always much below the other groups, suggesting that the relative intensity in routine jobs is much lower in this group with respect to the others, probably because the routinization process started earlier in the Nordic group.

When moving to the three separate task intensities, it is worth noting that the abstract one increases in basically all groups of countries, except in the Eastern group where it displays a non-monotonic trend. Further, differences in levels across groups emerge, suggesting that in Nordic and Anglo-Saxon

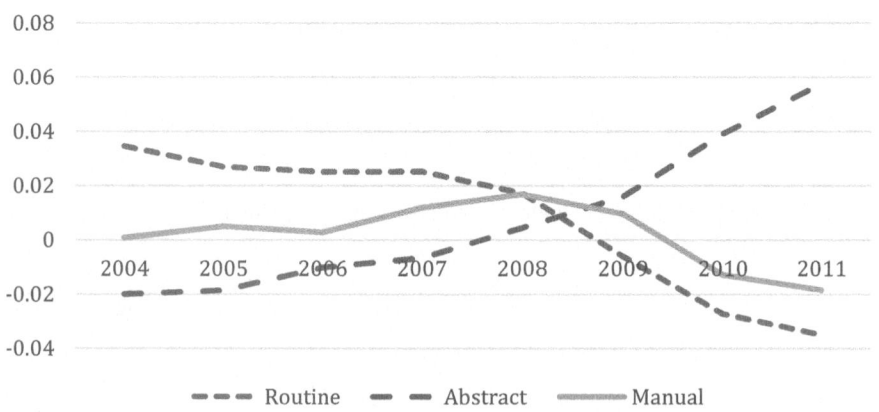

Fig. 4.2 Trends of the task intensity measures. All countries

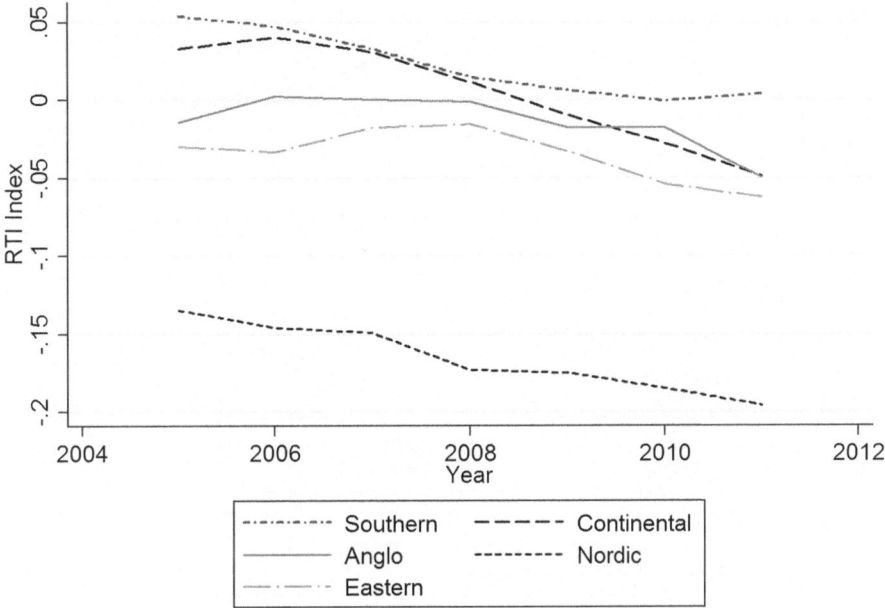

Fig. 4.3   RTI index in the country groups

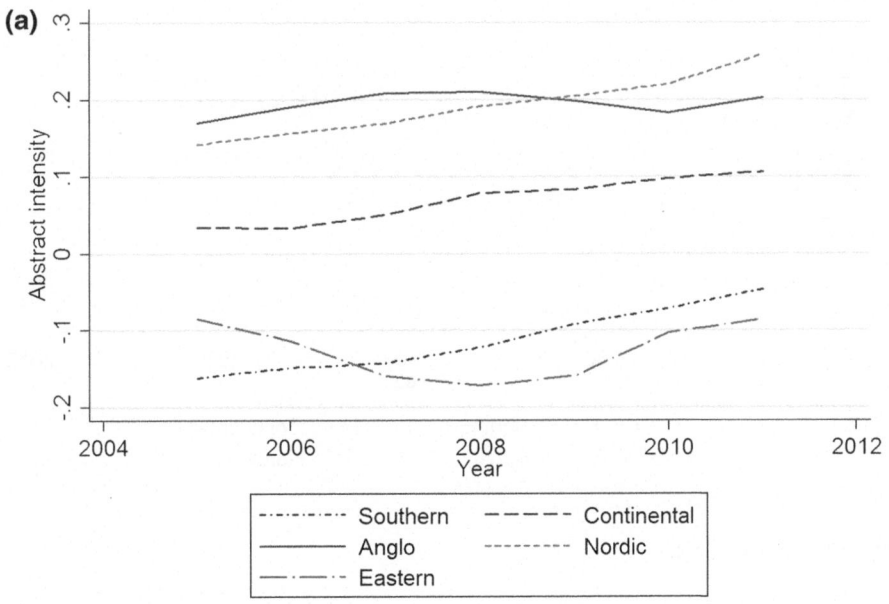

Fig. 4.4   Abstract (**a**), routine (**b**), and manual (**c**) intensities in the country groups

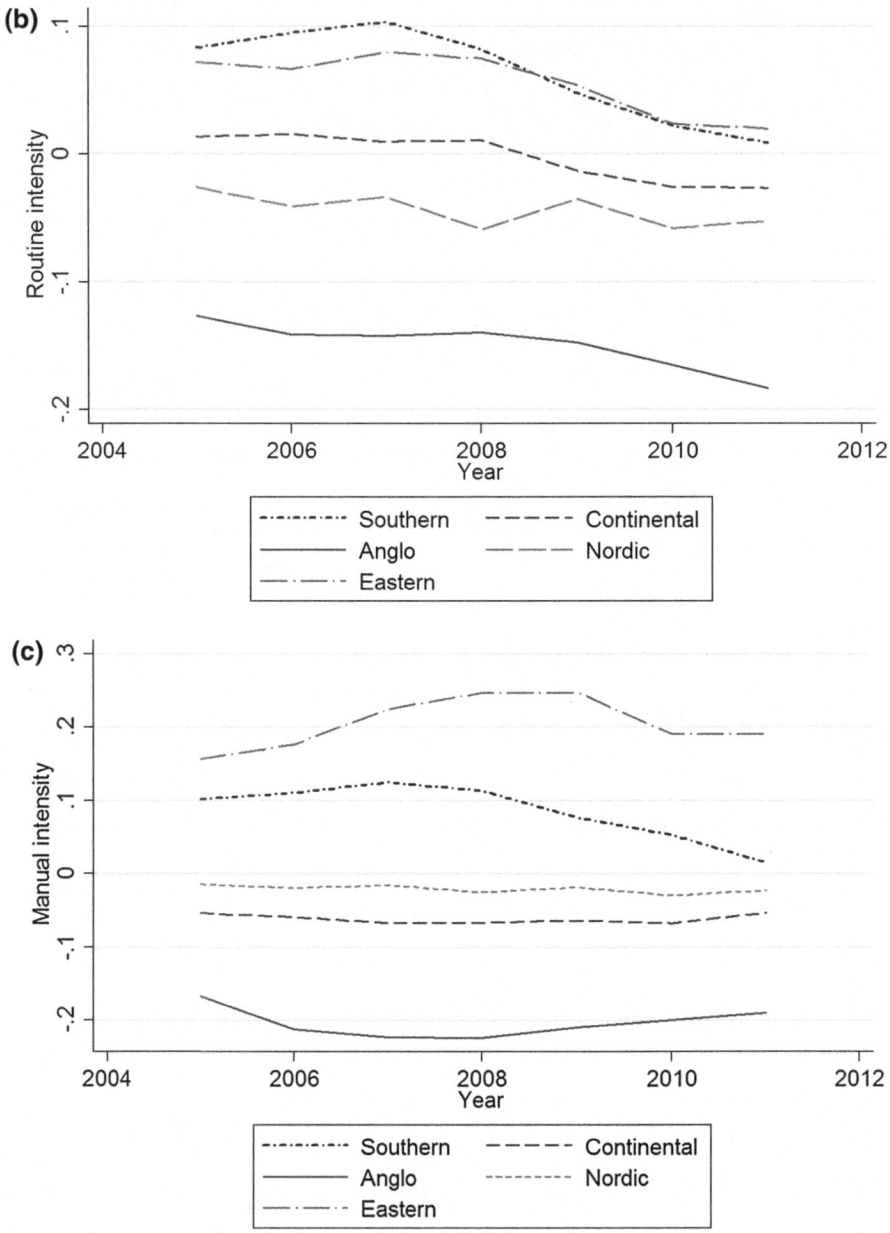

Fig. 4.4 (continued)

countries, the intensity of abstract jobs is higher, while it is lower in the Southern and the Eastern groups, with the Continental group being in the middle. As for the routine intensity, the trend is still the same (decreasing) in all groups. However, also in this case, differences in levels emerge, with Southern and Eastern having the highest intensity in routine jobs, and Nordic and Anglo-Saxon with the lowest.

More mixed is the evidence concerning the manual intensity. A stable trend is observed for the Continental and the Nordic groups, a decreasing trend for the Southern and for the Anglo-Saxon ones, and an increasing trend for the Eastern group.

Apart from the manual intensity, trends are rather similar across groups of countries, although differences in levels across groups seem to emerge, suggesting that the routinization process is at a different stage across groups.

4.3   The Determinants of Routinization

In this section, we focus on the identification of the determinants of routinization, in a 'descriptive' regression framework. In other words, we use as dependent variables our routinization measures, and we regress them, one at a time, on a set of socio-economic individual covariates (gender, age, education, whether single/married/no more married), some labour market covariates (permanent vs temporary job, full-time vs part-time job, hours worked). Since we make use of categorical variables, it is worth noting that the omitted group refers to individuals in the age class 55–64, single, with primary education). Moreover, all estimates include country dummies, to take into account time-invariant unobservable at the country level, and time dummies, to capture the business cycle.

Let us start by using the RTI index as dependent variable. In the first column of Table 4.1, the sample of all European countries is considered, while from column 2 to column 6, the regressions are carried out separately for the different country groups. The following findings emerge:

- Female workers display a higher incidence of RTI, in all columns.
- With respect to the omitted category (age 55–64), young individuals are characterized by the highest incidence of RTI. This correlation holds in all columns apart for the Anglo-Saxon group. In general, it emerges that RTI decreases along the career of the worker, and this is somehow unexpected, since one might believe that the new generations of workers are the ones with the highest level of education and more used, and more complement, to new technologies.

Table 4.1 Determinants of the RTI variable

	(1) All countries	(2) Southern	(3) Continental	(4) Anglo	(5) Nordic	(6) Eastern
Female	0.210***	0.205***	0.260***	0.273***	0.098***	0.219***
	(0.002)	(0.005)	(0.005)	(0.013)	(0.006)	(0.004)
Age 15–24	0.134***	0.186***	0.098***	0.032	0.079***	0.141***
	(0.005)	(0.011)	(0.012)	(0.027)	(0.014)	(0.009)
Age 25–34	0.104***	0.175***	0.059***	0.020	0.028***	0.114***
	(0.004)	(0.008)	(0.008)	(0.021)	(0.010)	(0.006)
Age 35–44	0.055***	0.121***	0.062***	−0.009	0.016*	0.018***
	(0.004)	(0.007)	(0.008)	(0.018)	(0.009)	(0.006)
Age 45–54	0.031***	0.091***	0.031***	−0.004	−0.001	0.007
	(0.003)	(0.007)	(0.008)	(0.018)	(0.009)	(0.006)
Upper secondary	0.052***	0.105***	0.060***	0.157***	−0.038***	−0.055***
	(0.003)	(0.005)	(0.006)	(0.017)	(0.009)	(0.006)
Tertiary	−0.578***	−0.508***	−0.571***	−0.425***	−0.637***	−0.700***
	(0.003)	(0.005)	(0.006)	(0.017)	(0.009)	(0.006)
Temporary	−0.073***	−0.143***	−0.103***	−0.070**	−0.087***	0.055***
	(0.003)	(0.005)	(0.007)	(0.029)	(0.010)	(0.006)
Part time	−0.012***	0.120***	−0.094***	−0.133***	−0.116***	−0.067***
	(0.004)	(0.009)	(0.008)	(0.021)	(0.011)	(0.010)
Married	−0.049***	−0.043***	−0.041***	−0.094***	−0.051***	−0.047***
	(0.003)	(0.006)	(0.006)	(0.015)	(0.007)	(0.005)
Divorced/separ/ widowed	−0.038***	−0.062***	−0.030***	−0.093***	−0.043***	−0.027***
	(0.004)	(0.010)	(0.009)	(0.022)	(0.010)	(0.007)
Hours worked	−0.002***	0.008***	−0.006***	−0.012***	−0.008***	−0.004***
	(0.000)	(0.000)	(0.000)	(0.001)	(0.000)	(0.000)
Constant	0.232***	−0.286***	0.426***	0.508***	0.330***	0.259***
	(0.010)	(0.019)	(0.018)	(0.046)	(0.024)	(0.023)
Country fixed effects	Yes	Yes	Yes	Yes	Yes	Yes
Year fixed effects	Yes	Yes	Yes	Yes	Yes	Yes
Observations	817,775	239,803	202,476	32,135	90,239	253,122
R-squared	0.099	0.067	0.102	0.108	0.121	0.109

Robust standard errors in parentheses
***$p < 0.01$, **$p < 0.05$, *$p < 0.1$

Table 4.2 Determinants of the routine intensity variable

Variables	(1) All countries	(2) Southern	(3) Continental	(4) Anglo	(5) Nordic	(6) Eastern
Female	−0.431***	−0.393***	−0.409***	−0.216***	−0.408***	−0.504***
	(0.002)	(0.004)	(0.005)	(0.012)	(0.007)	(0.004)
Age 15–24	0.140***	0.279***	0.252***	−0.007	0.002	0.023**
	(0.006)	(0.011)	(0.011)	(0.026)	(0.017)	(0.010)
Age 25–34	0.135***	0.246***	0.159***	0.067***	0.108***	0.052***
	(0.004)	(0.008)	(0.008)	(0.020)	(0.011)	(0.007)
Age 35–44	0.086***	0.155***	0.125***	0.040**	0.077***	0.010
	(0.004)	(0.007)	(0.007)	(0.017)	(0.009)	(0.007)
Age 45–54	0.055***	0.099***	0.083***	0.019	0.050***	0.015**
	(0.004)	(0.007)	(0.007)	(0.016)	(0.009)	(0.006)
Upper secondary	0.036***	−0.074***	0.108***	0.128***	0.040***	0.027***
	(0.003)	(0.005)	(0.006)	(0.016)	(0.010)	(0.007)
Tertiary	−0.379***	−0.301***	−0.316***	−0.154***	−0.295***	−0.556***
	(0.003)	(0.005)	(0.006)	(0.017)	(0.010)	(0.008)
Temporary	−0.080***	−0.110***	−0.109***	−0.029	−0.128***	−0.023***
	(0.003)	(0.005)	(0.007)	(0.024)	(0.011)	(0.006)
Part time	−0.145***	−0.080***	−0.161***	−0.248***	−0.256***	−0.297***
	(0.004)	(0.008)	(0.007)	(0.018)	(0.011)	(0.011)
Married	−0.015***	0.024***	−0.009*	−0.068***	−0.083***	−0.019***
	(0.003)	(0.005)	(0.005)	(0.015)	(0.008)	(0.006)
Divorced/separ/ widowed	−0.040***	−0.048***	−0.023***	−0.072***	−0.079***	−0.027***
	(0.004)	(0.008)	(0.008)	(0.020)	(0.011)	(0.008)
Hours worked	−0.001***	0.010***	−0.005***	−0.007***	−0.010***	−0.003***
	(0.000)	(0.000)	(0.000)	(0.001)	(0.000)	(0.000)
Constant	0.307***	−0.170***	0.361***	0.288***	0.529***	0.715***
	(0.010)	(0.019)	(0.017)	(0.045)	(0.027)	(0.028)
Country fixed effects	Yes	Yes	Yes	Yes	Yes	Yes
Year fixed effects	Yes	Yes	Yes	Yes	Yes	Yes
Observations	817,775	239,803	202,476	32,135	90,239	253,122
R-squared	0.106	0.097	0.098	0.041	0.091	0.140

Robust standard errors in parentheses

***$p < 0.01$, **$p < 0.05$, *$p < 0.1$

- With respect to the omitted category of primary education, workers with tertiary education display a lower incidence of RTI. Workers with upper secondary education display an intermediate value. These correlations apply to all groups of country, and are very robust from a statistical point of view.
- Permanent workers are associated to lower level of routinization, and this relation holds in all groups of countries. The same evidence applies for part-time workers.
- There is a positive correlation between hours worked and RTI, in all groups of countries.
- With respect to the omitted category (single), married and divorced/separated/widowed individuals have lower levels of RTI. Also, these relations apply to all groups of countries.

In Tables 4.2, we report as robustness check the same estimates with the routine intensity. It is actually reassuring that determinants have basically the same correlations as the ones detected in Table 4.1.

We do believe the most interesting evidence emerging from the analysis of the determinants concerns the fact that whatever routinization measure is used the young individuals display a higher level or routine jobs. In the following we will present some graphs showing the levels and trends of the RTI index for young individuals (15–34) and adults (over 34), for all groups of countries. It emerges clearly that for all groups, young individuals have a much higher level of RTI, and that for both groups, there is a decreasing trend. In particular, for the Southern group, the fall of the RTI index seems to be steeper than the one of the over 34, even if the difference between the two lines remain substantial even at the end of the period.

4.4   Is Routinization a Driver of Unemployment Inflows?

Another very debated issue in recent years is whether technological change fosters the destruction of jobs. There is an endless debate, with optimistic and pessimistic positions, started with David Ricardo (1817), and debated by well-known economists like John Maynard Keynes (1930) and Wassily Leontief (1983). Recently, the two bestseller books by Brynjolfsson and McAfee (2011, 2014), among others, have reintroduced this classic economic issue at the centre of the debate. Actually, there is little empirical evidence in recent years about the unemployment effect of technology (while more evidence is available for the composition issue), and the few existing papers provide mixed results.

The (scarce) existing literature generally uses aggregate data, at the country-industry level, in order to test whether some proxies of technological change affects the employment and unemployment variations. We do something different and somehow original in the literature for several reasons. First, we use as a proxy of technological change the routinization of an individual occupation. This represents a novelty that allows considering individual data instead of aggregate data, since for each individual in EU-SILC, it is possible to recover the occupation and then the associated measure of routinization. Second, since we make use of longitudinal data, we can follow individual careers over time, at least in the four years provided by the short EU-SILC panel. So far, the focus in the literature has been on changes of stocks, i.e., changes in employment share in each occupation (manual/service, routine, and abstract), without analysing the flows, i.e. individual longitudinal careers. Cortes (2016) and Cortes et al. (2016) point out that using longitudinal data matters: Flows are crucial to understand the mechanisms behind job polarization. In particular, Cortes (2016) shows for the United States that workers in routine jobs have higher probability to separate from their job. Interestingly, he finds that high skilled routine workers display higher probability to move to abstract jobs, while low ability routine workers are associated to higher probability to switch to manual jobs. Further, he shows that the decrease in routine employment is primarily due to the change in propensity of young workers to enter the labour market in routine jobs.

In such a framework, we follow individual careers over time in order to investigate the determinants for employed individuals to get into unemployment between time t and time $t+1$. In particular, we want to test whether this event is correlated to the routinization of the occupation before the job change. In the regressions, we introduce all the control variables that have been used in the analysis of the determinants of routinization (female, age classes, education, hours worked, permanent job, part time job, being married or no more married) as well as country dummies and year dummies.

We will consider two dependent variables. The first dependent variable is a dummy equal to one when an employed individual becomes unemployed the following year. The second dependent variable is equal to one when an employed individual become not employed (either unemployed or inactive) the following year. In this latter case, the dependent variable might be equal to one for individuals that exit voluntarily from the labour market, and for this reason, it might be considered as less reliable. Hence, the most reliable variable of interest concerns the transition from employment to unemployment.

Table 4.3 includes the related estimates when using the RTI index as routinization index. In column 1, we consider Europe as a whole. It emerges that the coefficient of RTI is positive and statistically significant, i.e., having a high RTI is positively associated to the probability to become unemployed the following year, even after controlling for observable differences in individual characteristics (mainly education and age), effort in the labour market (hours worked), selection into a permanent and a part time job, and marital status, as well as after controlling for unobservable differences across countries and time periods. The magnitude of the impact is not negligible: An increase of one standard deviation in RTI increases the baseline probability to become unemployed (2%) of 0.2%, i.e. of around 10%. This suggests that a new weak segment in the labour market can be identified. While in the last decades, the standard weak segment was always considered the one of unskilled labour in manual jobs, in recent years also individuals employed in routine occupations might be more likely to be exposed to higher risk of unemployment, even if they might not be necessarily unskilled workers.

When investigating this relation across different groups of countries, the positive coefficient of RTI holds in all groups except in the Anglo-Saxon one, suggesting that RTI is a driver on unemployment inflows, although groups of countries are characterized by very different institutional and industrial settings.

Table 4.4 refers to the same estimates when using as task variables the three task intensities (abstract, routine, and manual) that were collapsed in a synthetic measure when using the RTI index. The positive impact of routine intensity is now statistically significant only in two groups of countries, Southern and Continental, while it is not statistically different from zero (and actually very small in magnitude in the other groups). Interestingly, having high abstract intensity strongly reduces the probability to become unemployed, and this holds in all groups of countries. It is also worth noting that having high manual intensity is associated to a positive effect on the probability to become unemployed only in Europe as a whole and in the Southern group. This is on the one hand something unexpected, since manual tasks are usually associated to unskilled jobs, which are usually considered as the standard weak segment in the labour market. On the other, it is consistent with the polarization explanations, i.e. unskilled and manual jobs are increasingly demanded in the labour market.

Table 4.3 The impact of RTI on unemployment probabilities

Variables	(1) All countries	(2) Southern	(3) Continental	(4) Anglo	(5) Nordic	(6) Eastern
RTI	0.002***	0.001***	0.001***	0.001	0.002***	0.004***
	(0.000)	(0.000)	(0.000)	(0.000)	(0.001)	(0.000)
Female	−0.003***	−0.000	−0.003***	−0.006***	−0.005***	−0.004***
	(0.000)	(0.001)	(0.001)	(0.001)	(0.001)	(0.001)
Age 15–24	0.012***	0.022***	0.008***	0.010***	−0.005*	0.012***
	(0.001)	(0.003)	(0.002)	(0.003)	(0.003)	(0.002)
Age 25–34	0.004***	0.008***	0.005***	0.000	−0.006***	0.004***
	(0.001)	(0.001)	(0.001)	(0.002)	(0.001)	(0.001)
Age 35–44	0.002***	0.002*	0.002**	−0.001	−0.003***	0.003***
	(0.001)	(0.001)	(0.001)	(0.002)	(0.001)	(0.001)
Age 45–54	0.001	−0.002	−0.001	−0.001	−0.002*	0.005***
	(0.001)	(0.001)	(0.001)	(0.002)	(0.001)	(0.001)
Upper secondary	−0.015***	−0.017***	−0.012***	−0.008***	−0.006***	−0.026***
	(0.001)	(0.001)	(0.001)	(0.002)	(0.001)	(0.002)
Tertiary	−0.026***	−0.029***	−0.018***	−0.010***	−0.012***	−0.041***
	(0.001)	(0.001)	(0.001)	(0.002)	(0.002)	(0.002)
Temporary	0.054***	0.062***	0.056***	0.014***	0.041***	0.046***
	(0.001)	(0.001)	(0.002)	(0.004)	(0.003)	(0.002)
Part time	0.009***	0.014***	0.005***	0.002	0.007***	0.006**
	(0.001)	(0.002)	(0.001)	(0.002)	(0.002)	(0.002)
Married	−0.006***	−0.007***	−0.005***	−0.003**	−0.009***	−0.005***
	(0.000)	(0.001)	(0.001)	(0.002)	(0.001)	(0.001)
Divorced/separ/ widowed	0.001*	−0.002	0.005***	−0.000	−0.001	0.002
	(0.001)	(0.002)	(0.001)	(0.002)	(0.002)	(0.001)
Hours worked	0.000***	0.000***	−0.000**	−0.000**	0.000	−0.000
	(0.000)	(0.000)	(0.000)	(0.000)	(0.000)	(0.000)
Constant	0.025***	0.021***	0.032***	0.025***	0.034***	0.061***
	(0.002)	(0.003)	(0.003)	(0.005)	(0.004)	(0.004)
Country fixed effects	Yes	Yes	Yes	Yes	Yes	Yes
Year fixed effects	Yes	Yes	Yes	Yes	Yes	Yes
Observations	817,775	239,803	202,476	32,135	90,239	253,122
R-squared	0.027	0.035	0.024	0.007	0.016	0.022

Robust standard errors in parentheses
***$p < 0.01$, **$p < 0.05$, *$p < 0.1$

Table 4.4 The impact of task intensities on unemployment probabilities

Variables	(1) All countries	(2) Southern	(3) Continental	(4) Anglo	(5) Nordic	(6) Eastern
Routine intensity	0.000	0.001**	−0.002***	−0.001	0.000	0.000
	(0.000)	(0.000)	(0.000)	(0.001)	(0.000)	(0.000)
Abstract intensity	−0.006***	−0.006***	−0.005***	−0.002***	−0.004***	−0.008***
	(0.000)	(0.001)	(0.000)	(0.001)	(0.001)	(0.000)
Manual intensity	0.001***	0.003***	0.001	−0.001	−0.000	0.001
	(0.000)	(0.000)	(0.000)	(0.001)	(0.001)	(0.000)
Female	−0.002***	0.002**	−0.003***	−0.007***	−0.005***	−0.002***
	(0.000)	(0.001)	(0.001)	(0.001)	(0.001)	(0.001)
Age 15–24	0.011***	0.021***	0.007***	0.010***	−0.006**	0.012***
	(0.001)	(0.003)	(0.002)	(0.003)	(0.003)	(0.002)
Age 25–34	0.004***	0.006***	0.004***	0.000	−0.006***	0.004***
	(0.001)	(0.001)	(0.001)	(0.002)	(0.001)	(0.001)
Age 35–44	0.001**	0.001	0.002*	−0.001	−0.003***	0.003**
	(0.001)	(0.001)	(0.001)	(0.002)	(0.001)	(0.001)
Age 45–54	0.000	−0.002*	−0.001	−0.001	−0.002*	0.005***
	(0.001)	(0.001)	(0.001)	(0.002)	(0.001)	(0.001)
Upper secondary	−0.013***	−0.013***	−0.009***	−0.007***	−0.006***	−0.023***
	(0.001)	(0.001)	(0.001)	(0.002)	(0.001)	(0.002)
Tertiary	−0.017***	−0.018***	−0.012***	−0.009***	−0.009***	−0.028***
	(0.001)	(0.001)	(0.001)	(0.002)	(0.002)	(0.002)
Temporary	0.053***	0.060***	0.056***	0.014***	0.041***	0.044***
	(0.001)	(0.001)	(0.002)	(0.004)	(0.003)	(0.002)
Part time	0.007***	0.013***	0.005***	0.002	0.007***	0.003
	(0.001)	(0.002)	(0.001)	(0.002)	(0.002)	(0.002)
Married	−0.006***	−0.007***	−0.005***	−0.003**	−0.008***	−0.005***
	(0.000)	(0.001)	(0.001)	(0.002)	(0.001)	(0.001)

Table 4.4 (Continued)

Variables	(1) All countries	(2) Southern	(3) Continental	(4) Anglo	(5) Nordic	(6) Eastern
Divorced/separ/widowed	0.001*	−0.002	0.005***	−0.000	−0.001	0.002
	(0.001)	(0.002)	(0.001)	(0.002)	(0.002)	(0.001)
Hours worked	0.000**	0.000***	−0.000	−0.000**	0.000	−0.000**
	(0.000)	(0.000)	(0.000)	(0.000)	(0.000)	(0.000)
Constant	0.021***	0.018***	0.028***	0.024***	0.033***	0.057***
	(0.002)	(0.003)	(0.003)	(0.005)	(0.004)	(0.004)
Country fixed effects	Yes	Yes	Yes	Yes	Yes	Yes
Year fixed effects	Yes	Yes	Yes	Yes	Yes	Yes
Observations	817,775	239,803	202,476	32,135	90,239	253,122
R-squared	0.028	0.036	0.024	0.007	0.016	0.023

Robust standard errors in parentheses
***$p < 0.01$, **$p < 0.05$, *$p < 0.1$

4.5   Conclusion

In this chapter, we provide evidence on the ongoing routinization processes in Europe. We derive the following evidence: Routinization processes do not depend on the type of variable considered; groups of countries characterized by very different institutions, cultures, and labour market conditions, share similar routinization trends; routinization levels are different across groups, suggesting that groups of countries can be placed at different stage in the technological/routinization process; routinization seems to represent a driver, among others, of unemployment inflows: Individuals in routine jobs display, ceteris paribus, an higher probability to become unemployed. Further, the magnitude of the effect is not negligible: A one standard deviation increase in the RTI index entails a 10% increase of getting into unemployment. This evidence challenges the view that only unskilled workers represent the weak segment in the labour market, as it was in the last decades. This evidence applies also to routine workers, that are not necessarily unskilled. If confirmed, this issue would represent an important new dimension of the policy debate in the next years, to think about some new targeting dimensions of the active and passive labour market policies, with two main objectives. On the one hand, to ease the reallocation of routine workers towards other types of jobs, and on the other hand to provide some sort of additional institutional insurance for workers more exposed to unemployment risks.

Furthermore, future research is needed to investigate in more details what there is behind these findings. In particular, on the one hand, this result might be due to a casual effect of having a routine job on unemployment inflows. On the other hand, this effect might be due to a self-selection effect into routine jobs. In other words, it might be the case that unskilled individuals self-select into routine jobs, and hence the effect on unemployment inflows might be actually driven by this self-selection and not by a causal impact of performing a routine task. These issues should be analysed by using a counterfactual analysis, controlling for unobserved heterogeneity and exploiting some form of exogenous variation in order to identify from an econometric point of view the causal effect of having a routine job.

References

Autor, David H., and David Dorn. 2013. "The Growth of Low-Skill Service Jobs and the Polarization of the US Labor Market." *American Economic Review* 103 (5): 1553–1597.

Autor, David, Frank Levy, and Richard Murnane. 2003. "The Skill Content of Recent Technological Change: An Empirical Exploration." *The Quarterly Journal of Economics* 118 (4): 1279–1333.

Autor, David, Lawrence Katz, and Melissa Kearney. 2006. "The Polarization of U.S. Labor Market." *American Economic Review* 96 (2): 184–194.

Brynjolfsson, Erik, and Andrew McAfee. 2011. *Race Against the Machine: How the Digital Revolution is Accelerating Innovation, Driving Productivity, and Irreversibly Transforming Employment and the Economy.* Lexington, MA: Digital Frontier Press.

Brynjolfsson, Erik, and Andrew McAfee. 2014. *The Second Machine Age: Work, Progress and Prosperity in a Time of Brilliant Technologies.* New York: W. W. Norton.

Cortes, Mathias. 2016. "Where Have the Middle-Wage Workers Gone? A Study of Polarization Using Panel Data." *Journal Of Labour Economics* 34 (1): 63–105.

Cortes, Mathias, Nir Jaimovich, Christopher Nekarda, and Henry Siu. 2016. *The Micro and Macro of Disappearing Routine Jobs: A Flows Approach.* New York: Mimeo.

Engel, Melissa, and Sandra Schaffner. 2012. "How to Use the EU-SILC Panel to Analyse Monthly and Hourly Wages." Ruhr Economic Paper 336, Rheinisch-Westfälisches Institut für Wirtschaftsforschung.

Frick, Joachim R., and Kristina Krell. 2010. "Measuring Income in Household Panel Surveys for Germany: A Comparison of EU-SILC and SOEP." SOEP papers on Multidisciplinary Panel Data Research 265.

Goos, Maarten, Alan Manning, and Anna Salomons. 2009. "Job Polarization in Europe." *American Economic Review* 99 (2): 58–63.

Hauser, Richard. 2008. "Problems of the German Contribution to EU-SILC: A Research Perspective, Comparing EU-SILC, Microcensus and SOEP." German Council for Social and Economic Data (RatSWD) 80.

Keynes, John Maynard. 1930. *Economic Possibilities for Our Grandchildren.* New York: MIT Press.

Leontief, Wassily. 1983. National Perspective: The Definition of Problems and Opportunities. In *National Academy of Engineering Symposium, The Long-Term Impact of Technology on Employment and Unemployment.* Washington, DC: National Academy Press.

Ricardo, David. 1817. *On the Principles of Political Economy and Taxation.* London: Dent & Sons.

5

Labor Markets in the Digital Economy: Modeling Employment from the Bottom-Up

Jonathan Liebenau

5.1 Introduction

The effects upon labor markets of changes in the digital economy are poorly understood because analyses at the appropriate levels of jobs within specified sectors has been inadequately studied. Contrasting popular views can be characterized as either neo-Luddite—technology destroys employment or at least meaningful work, or techno-utopian—unemployment will be eliminated by massive demand for a wide variety of high-quality work. Despite a constant flow of expert reports promoting each of these positions, I will show here how a bottom-up accounting of both job destruction and job creation provides a much more accurate assessment of the effects of investment in the digital economy. While the effect differs very significantly from sector to sector and place to place, overall we see very modest but constant growth in most places where investments in the digital economy are robust.

This is important both for the general understanding of the effects of the digital economy and for our ability to be specific about the impacts of investments of certain kinds upon local economies. The problem of making

I acknowledge the collaboration of Patrik Karrberg of the LSE worked with me on three of the studies upon which this chapter is based, Michael Mandel of the Progressive Policy Institute on another, and Robert B. Atkinson, Daniel Castro and Stephen Ezell all of the Information Technology and Innovation Foundation on others.

J. Liebenau (✉)
Department of Management, London School of Economics, London, UK

© The Author(s) 2018
L. Pupillo et al. (eds.), *Digitized Labor*, https://doi.org/10.1007/978-3-319-78420-5_5

sense of employment effects is shrouded by controversy about job creation and job losses. On the one side, we aspire to find opportunities to alleviate unnecessary unemployment and to find good quality jobs for all those who have invested in attaining skills. On the other side, we feel compelled to find ways to drive productivity growth even where firms are unable to increase output proportionately while simultaneously employing more staff.

From either position, it is important to have a clear idea of labor market skills requirements for the sake of education and training policies, if only to make sense of the constant refrain coming from employers and amplified by researchers who perennially claim that the labor market is desperately in need of hundreds of thousands of newly skilled workers. There is a constant refrain of laments about skills shortages that goes back decades.[1] What, exactly, the labor market was missing changed constantly, so the supposed dearth of hardware designers of the past was displaced by frenzied demands that hundreds of thousands of software engineers be trained, followed by confident predictions that the economy's potential for growth is hampered by the lack of sufficient data analysts, etc. These claims are all belied by two clear trends: The number of digital economy workers is only gradually rising, and unemployment among information and communications skilled workers is almost constantly half of overall unemployment in OECD economies. All the while the overall unemployment rate fluctuates very significantly within about 20-year periods.

So, we see a simple contrast between our ambition to solve unemployment problems while we promote productivity growth and "modernization" which at the firm level often means displacing large numbers of lesser skilled workers with either smaller numbers of higher skilled workers or better technologies, or both.

Most of the discussion to date about the effects of the digital economy has been based on one of two different kinds of analysis. There are those, such as Robert Gordon and John van Reenan, who start from the productivity growth trends in the macroeconomy and project broad labor futures, and there are those, such as Eric Brynjolfsson and Andrew McAfee (2011), who start from the affordances of the technology and imagine what work might be. The first of these either concludes that the growth in demand for knowledge workers will be so great as to provide unlimited high-quality employment. Or, as Robert Gordon does, regards the recent boom in productivity to be anomalous in historical trends and forecasts slow growth with no significant employment consequences.[2] Those who focus on technology's affordances similarly split between optimists who believe that the promise of each new technology boom in the digital economy, currently big data analytics and the "internet of things", will soak up unemployment.[3] Or they become pessimists, such as Brynjolfsson, who write about the robotics of the near future, both physical and virtual, that will displace vast swathes of meaningful work.

The work I report on here, based on four studies on labor market effects of information and communication technologies (ICT)[4] takes a very different approach. While my interests are, similarly, on the character of work and the large-scale effects of ICT upon employment trends, I start from two solid pillars to model trends. The first is rooted in the existing long-term technology investments and policy commitments, seen in governments, large firms, and social trends. These include the manner in which smartphone industries work, investments in smart energy and the shift currently underway in areas we see such as in the fintech firms and in logistics, retailing, and interconnections. The second concerns the specifics of job creation and displacement; i.e., the ways in which tasks are performed, the people employed to perform them, the cost of engaging such employees, and, crucially, the effects that has on the labor market more generally. I also make use, in a distinctly conservative manner, of employment multipliers. While this remains a somewhat controversial technique, it has been used in studies both in the econometric literature on labor markets (Etro 2009)[5] and in the applied and policy literature (Kapstein 2008).[6]

5.2   Methodology and Hypotheses

The starting point of this analysis is the relationship between investments and the character of specific tasks. The input is the presumed spending, by governments or private sector investors, in known digital economy organizations. From this, we can get a first estimate of job creation and displacement. We must take into account the technologies themselves to capture the real changes that are likely to have effects upon organizational productivity, including details of what components are used, how production speed might be affected, whether production processes including tasks such as producing lines of code are enhanced, etc. This bottom-up approach emphasizes the specificity of what firms are doing, their labor market catchment and other features of their location and takes us beyond NIC categories and associated labor statistics. This is because among the key factors determining the effects of investments are the extent to which intermediate markets are filled by domestic suppliers as opposed to imports, the extent to which national procurement policies might affect local sourcing, and which industrial policies are likely to have effect, especially with regard to incentives offered to specific sectors or for certain categories of investment as has long been the case with R&D spending.

In summary, our analysis takes a bottom-up approach that focuses on industry sector level within national contexts. While we use some survey results to help in estimating trends in ICT staff redeployment[7] and evidence about national procurement and infrastructure development priorities, most of the evidence comes from data about labor market composition in relation to product and services development trends.

Sources include national statistical bodies and industry association data, International Labour Organization and International Telecommunications Union statistics, evidence about the productivity of key choices such as in-house versus cloud-based activities, and descriptions of job types and industry structure. In these ways, we can build a rounded picture of the character of the labor market, the distribution of activity between large versus small and medium firms, the effects of trade on the capacity of the domestic industry to respond, and related features of the employment landscape. The strengths and weaknesses of each of these sources of evidence are specified below, within the context of each of the four main studies reported.

Study topic	Research questions	Sectors	Evidence
UK recovery	What effect will accelerated spending on digital economy projects have on employment?	Broadband upgrading; intelligent transport systems; smart energy networks	Domestic market capabilities; industry structure; multiplier effects; labor market data
Modelling the cloud (1)	How do existing cloud services trends in exemplary sectors affect job losses and gains in Germany, Italy, Britain & the USA?	Aerospace; smartphone services	Sectoral features & trends; domestic conditions for cloud utilization
Modelling the cloud (2)	How might projected trends in cloud services affect the Turkish labor market and what does that indicate about emerging economy contexts?	Automotive manufacturing; smartphone services	Sectoral features & trends; domestic conditions for cloud utilization

Study topic	Research questions	Sectors	Evidence
Digital London	What characterizes and explains London's tech/info labor market since the financial crisis of 2008?	All areas of info/tech employment; special focus on big data & fintech jobs	Job advertisements; geographically linked labor market data

Four employment effects studies

While each of the main studies referred to addresses specific research questions concerning the conditions of labor markets for specific sectors in particular countries, in general, we are able to address four main issues. (1) The first addresses the question that, given existing investment intentions, how can an accelerated timescale boost sustainable employment? (2) The second tests the assumption that domestic capabilities along with sourcing policies make significant differences in labor trends. (3) Next, we test the extent to which different technologies respond differently because of mix of job type. (4) Finally, we account for the distribution of effects between large versus small and medium size firms.[8]

All of these concerns cut across each of the four studies under consideration, which are presented in chronological order. In the final section, we will consider what the syntheses of these findings reveal. The first is a study of UK investment effects in broadband upgrading, intelligent transport systems, and smart energy technologies. The second and third address questions about the employment effects of cloud computing trends in international comparison, taking into account both manufacturing and services sector effects. The fourth uses technology and information sector employment data for London in a study that allows for direct contrasts with that in New York and the San Francisco/Silicon Valley regions.

5.3   UK Investment in Recovery[9]

In order to study the impact of certain kinds of digital economy investments in the UK, we selected three cases that fit the main criteria of being potentially high impact infrastructure developments that had long been identified as public policy priorities. The fact that they had been seen as important areas for development meant that there were few policy or planning restrictions to worry about; these were not "pie-in-the-sky" ambitions but gradual, if significant, extensions of existing transportation, energy, and broadband policies. The three specific cases were in intelligent transportation systems, smart energy technologies, and broadband upgrade. For each of these, we postulated £5 billion of spending spread over around three years.

Our research questions were threefold:

1. Given existing investment intentions, what affect will accelerated timescales have on sustainable employment?
2. To what extent do domestic capabilities and sourcing practices make a significant difference?
3. To what extent do differences among investments in different technologies affect the mix of job types?

The first of these was intended to address whether short-term boosts to employment are more or less likely to be sustainable where digital investments are made as opposed to other, known ephemeral effects of literally "shovel-ready" construction projects.[10] As was pointed out repeatedly during the initial discussions of the so-called "Obama stimulus plan", unemployment can most quickly be ameliorated by providing infrastructure construction jobs where no significant preparation is needed, as would be the case with road resurfacing or bridge repairs.

The second focused on production and trade data and concerned the capacity of domestic producers to fulfill the requirements of the stated scale of expansion.

We were able to calculate the specific domestic employment effects taking into account the time periods of the projects, the existing trade patterns, and the extent to which the intermediate markets for components used in, for example smart energy meters, is well developed. This requires a rather detailed analysis of the content of the technologies and an assessment of the qualities of the related software industries, component manufacturing, and the ability to deploy such products and services quickly.

The third question is addressed in part by considering the multiplier effects of investment in particular kinds of jobs. This takes into account, for example, the differences between construction work, which has high employment multiples because of the knock-on effects for materials and machinery suppliers, and software development, which has rather minimal multiplier effects that take into account only what happens with, for example, office space and supplies and directly related services (such as lunch restaurants).

Since the analysis closely parallels a slightly earlier study of the employment effects in the United States (Atkinson et al. 2009), our findings show interesting contrasts in some specific areas. One is that the domestic capabilities in the UK are surprisingly good, with significant capabilities not only in electricity grid production but also in its ability to produce smart meters. Similarly, for intelligent transport systems, UK small and medium businesses are well placed to take advantage of the considerable investment necessary for infrastructure construction including physical networks installed in roadways as well as the production of electronics for vehicles. The comparison with regard to broadband networks, where much of the technology is likely to be imported from either the United States or China, shows less of a disproportionate effect on small and medium sized enterprises but a significant employment effect overall, especially for the larger firms that dominate the telecommunications sector.

The overall effects, as seen in Table 5.1, are revealed to be even more distinct when comparing sectors, as Tables 5.2 and 5.3 show. Since the structures of the industries in terms of the split between large and smaller firms differs, and since the investments imply differing levels of investment in jobs that are associated with high versus low employment multipliers, the contrast is important. We will return to this later when we consider how policy priorities, for the types as well as locations of jobs created, might be made.

Table 5.1 Estimates of UK jobs created or retained by investment in network infrastructures for 1 year

ICT investment	Investment	Total jobs	Small business jobs
Broadband networks	£5 billion	280,500	94,000
ITS	£5 billion	188,500	120,000
Smart power grid	£5 billion	231,000	146,000
Total	£15 billion	700,000	360,000

Table 5.2 UK jobs created or retained for 1 year by a £5 billion broadband investment

Job type	Total jobs	Small business jobs
Direct	76,500	22,500
Indirect and induced	134,500	37,000
Network effect	69,500	34,500
Total jobs	280,500	94,000

Table 5.3 UK jobs created or retained for 1 year by a £5 billion its investment

Job type	Total jobs	Small business jobs
Direct	62,500	44,000
Indirect and induced	79,000	53,000
Network effect	47,000	23,000
Total jobs	188,500	120,000

This study demonstrated the differential effects of spending on different digital economy sectors and the differences one might expect among direct and indirect job creation. It is also clear that while some jobs are quickly created because of the direct effect of employment practices, others lag until, for example, materials purchases are placed. Construction jobs, which account for a considerable proportion of any infrastructure upgrading project, occur early in planning and fade quickly, while many indirectly stimulated jobs appear later and some are sustained.

5.4   Employment Effects of Cloud Services

The second and third studies were conducted using a very similar methodology but focusing not on the broad effects of single, large investments but rather on one set of digital economy services. To understand what impact the move to cloud computing might have, one needs again to consider the relationship between investment trends as well as the specifics of job characteristics for those engaged in various aspects of cloud services work. This includes the work of those developing, marketing and implementing new software for such services, but also crucially the production and maintenance of data centers. In addition, the new skills needed by the users of such services are going to require hiring as the relevant tasks shift from the in-house provision of data analysis to the utilization of outputs provided by cloud services providers. The major consideration in this analysis, however, is that we can also closely estimate the unemployment effects of cloud services as users divest themselves of hardware and staff formerly devoted to providing in-house services. We were strongly influenced in estimating the number of people likely to be made unemployed by this shift by the results of an interview-based survey of chief information officers and other executives of large enterprises in many countries (referred to in footnote 8, above). The majority of respondents claimed that they would take advantage of most of the productivity gains associated with cloud services not by letting their skilled ICT employees go, but rather by redeploying them. While we estimate that there will be some unemployment generated, overall if the rate of spending on cloud services continues as per the current trend, the effect is positive, if modest.

Our research questions were again threefold:

1. Given existing investment intentions, what affect will the growth of cloud services have on sustainable employment?
2. To what extent do sectoral distinctions make a significant difference?

3. What effects might we expect from changes in policies such as those that concern transborder data flow and energy pricing?

For the first of these studies, focusing on employment effects in the UK, USA, Germany, and Italy, we chose three sectors, aerospace (exemplifying mature manufacturing), smartphone services (exemplifying cloud-based business), and mobile operators (an element of infrastructure effects). The second focused only on Turkey, which allowed us to delve deeper into policy effects as well as showed some interesting distinctions relevant for emerging market conditions. That study repeated the focus on smartphone services and mobile operators, but substituted automobile manufacturing for aerospace. For both studies, we used industry and company sources for both suppliers of cloud services as well as customers to understand spending and investment trends, and disaggregated the elements of each industry that utilize cloud services.[11]

Our methodology again rested on a mix of analyses of domestic information and communication technology capabilities and trends, education as it affects both management and the technical workforce, and energy policy. Once again we used our own large-scale interview evidence of large enterprise chief information officers to gauge employers' intentions with regard to the redeployment versus reduction of staffing (Table 5.4, Fig. 5.1).

We repeated this form of analysis for each of the specific configurations of technology under analysis. Taking into account the cost breakdown of data centers, we were able to estimate the cloud jobs of different components for each of the sectors assessed. For the smartphone service sector, this added up to considerable growth during the period 2010–2014, from under 20,000 jobs to almost 55,000 (Table 5.5).

Table 5.4 Jobs

	2010 IT Jobs (direct, indirect, and induced)	2014 IT Jobs (direct, indirect, and induced)	2010 Direct cloud jobs	2014 Direct cloud jobs
US Aerospace	112,000	128,000	1760	2770
US SPS	84,000	148,000	2060	5210
UK Aerospace	11,200	12,100	210	320
UK SPS	5890	13,990	200	870
Ger Aerospace	16,200	17,800	290	420
Ger SPS	8840	16,840	230	800
It Aerospace	3560	3900	60	100
It SPS	6950	12,730	200	700

SPS smartphone services

For the purposes of our analysis, we devised the following structure for our smartphone model:

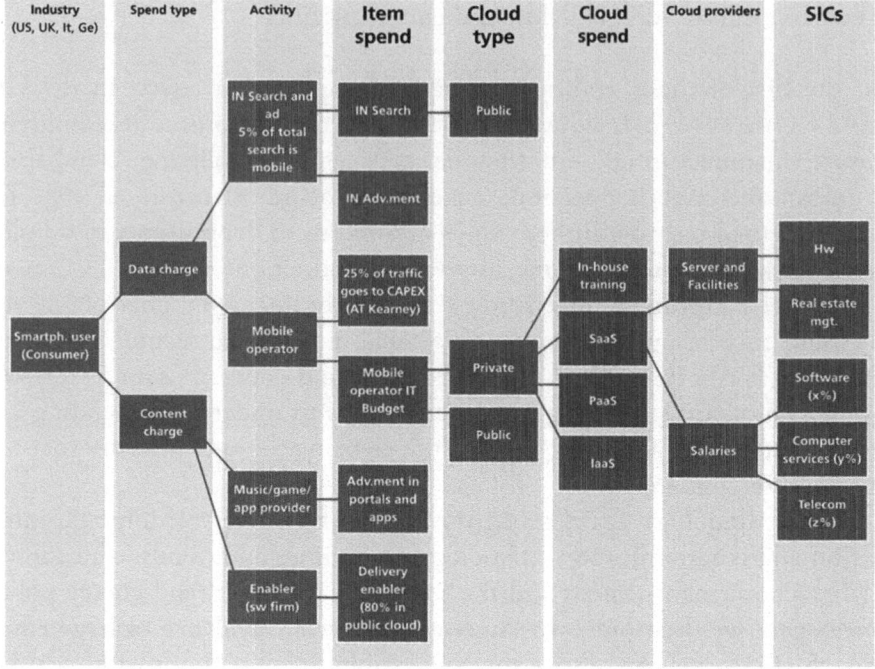

Fig. 5.1 Smartphone model (*Note* the figure is a simplified view whereas in the model all item spends are broken down and analysed into cloud type and eventually SICs)

Table 5.5 Overview of the US smartphone service sector (*Sources* Juniper Research, AT Kearney, HSBC, IDC, Corporate reports (2011))

United States	2010	2014	Cloud Jobs 2014
Smartphone service turn-over	$27.8bn	$46.5bn	54,500
Data revenues (non-messaging)	$27bn	$42.5bn	31,400
Enterprise Applications	$248m	$1.2bn	6770
Consumer Applications (subs)	$23m	$478m	2720
Consumer Apps (advertisement)	$16m	$138m	790
Mobile Search (advertisement)	$503m	$2.25bn	12,780
Related US cloud jobs	19,600	54,500	
SME Cloud spending as share of (total)			
Public cloud	55%	63%	
Private cloud	22%	27%	

A similar growth trend can be seen for the US aerospace sector, but with the important difference in the structure of the industry where growth as a result of small- and medium-sized enterprise spending on cloud services accounting for a much higher proportion of the smartphone services business (because of the large number of apps developers engaged), versus the large firm-dominated character of the aerospace sector (Table 5.6).

The pattern is similar in the UK, with a slightly larger proportion of the industry structure comprised of SMEs, but because of the different weightings placed on segments of the industry, slightly different growth rates, also (Table 5.7).

A much larger difference in the share of spending on cloud services from aerospace SMEs is apparent in the UK industry, indicative of a much higher reliance on a more diversified supply chain there than in the United States (Tables 5.8, 5.9, 5.10).

Table 5.6 Overview of the US aerospace sector

	2010	2014
Aerospace service revenues	$210bn	$241bn
Total induced, indirect, and direct jobs	15,400	24,200
SME Cloud spending as share of (total)		
Public cloud	7%	8%
Private cloud	3%	4%

Table 5.7 UK (*Sources* Juniper Research, AT Kearney, HSBC, IDC, Corporate reports (2011))

UK	2010	2014	Cloud jobs 2014
Smartphone service turn-over	£1.6bn ($2.5bn)	£3.9bn ($6.1bn)	4040
Data revenues (non-messaging)	£1.5bn	£3.3bn	2144
Enterprise Applications	£47m	£235m	787
Consumer Applications (subs)	£2.8m	£73m	246
Consumer Apps (advertisement)	£2.2m	£21m	72
Mobile Search (advertisement)	£26m	£236m	791
Total induced, indirect, and direct jobs	900	4040	
SME Cloud spending as share of (total)			
Public cloud	57%	63%	
Private cloud	27%	33%	

Table 5.8 UK

UK	2010	2014
Aerospace revenues	£22.2bn ($34.7bn)	£24.3bn ($37.8bn)
Total induced, indirect, and direct jobs	880	1340
SME Cloud spending as share of (total)		
Public cloud	30%	33%
Private cloud	16%	18%

Table 5.9 Germany

Panel A			
Data center cost breakdown	Public cloud (%)	Private cloud (%)	Internal data center (%)
Unit costs (per M $)			
Facilities/infrastructure	29.2	21.9	15.8
IT hardware (annualized)	42.1	31.7	22.8
Software	0.0	2.8	2.5
Electricity costs	17.7	20.7	28.6
Network fees	1.8	1.4	1.0
Property taxes	2.7	2.0	1.4
Staff	6.5	19.4	27.9
IT administrators	1.8	15.9	25.3
Facilities site management	1.8	1.4	1.0
Maintenance	0.9	0.7	0.5
Janitorial and landscaping	0.6	0.5	0.3
Security	1.4	1.0	0.7

Panel B			
Germany	2010	2014	Related jobs 2014
GER smartphone service turn-over	€3.1bn ($4.3bn)	€4.8bn ($6.5bn)	4840
Data revenues (non-messaging)	€2.2bn	€4.2bn	2520
Enterprise applications	€57m	€266m	970
Consumer applications (subs)	€3.3m	€83m	300
Consumer apps (advertisement)	€2.6m	€24m	90
Mobile search (advertisement)	€30m	€267m	970
Total induced, indirect, and direct jobs	1273	4840	
SME cloud spending as share of (total)			
Public cloud	54%	62%	
Private cloud	25%	32%	

Panel C		
Germany	2010	2014
Aerospace revenues	€24.7bn ($34bn)	€27.2bn ($37.6bn)
Total induced, indirect, and direct jobs	1490	2100
SME cloud spending as share of (total)		
Public cloud	5%	6%
Private cloud	2%	3%

Panel A: Germany has relatively higher electricity costs and lower staffing costs than the UK and the USA

Panel B: *Sources* Juniper Research, AT Kearney, HSBC, IDC, Corporate reports (2011)

Table 5.10   Italy

Panel A

Data center cost breakdown	Public cloud (%)	Private cloud (%)	Internal data center (%)
Unit costs (per M $)			
Facilities/infrastructure	27.9	20.9	15.0
IT hardware (annualized)	40.3	30.2	21.7
Software	0.0	2.7	2.4
Electricity costs	21.9	26.4	34.8
Network fees	1.8	1.3	0.9
Property taxes	2.5	1.9	1.4
Staff	5.6	16.6	23.8
IT administrators	1.9	16.9	27.0
Facilities site management	2.0	1.5	1.1
Maintenance	1.0	0.7	0.5
Janitorial and landscaping	0.7	0.5	0.4
Security	1.5	1.1	0.8

Panel B

Italy	2010	2014	Related Jobs 2014
GER smartphone service turn-over	€2.1bn ($2.9bn)	€3.9bn ($5.4bn)	3450
Data revenues (non-messaging)	€2bn	€3.37bn	1790
Enterprise applications	€40m	€199m	600
Consumer applications (subs)	€2.4m	€62m	190
Consumer apps (advertisement)	€1.9m	€18m	60
Mobile search (advertisement)	€31m	€277m	820
Total induced, indirect, and direct jobs	937	3450	
SME cloud spending as share of (total)			
Public cloud	54%	63%	
Private cloud	25%	33%	

Panel C

Italy	2010	2014
Aerospace revenues	€8bn ($11bn)	€8.8bn ($12.2bn)
Total induced, indirect, and direct jobs	280	380
SME cloud spending as share of (total)		
Public cloud	30%	32%
Private cloud	16%	18%

Panel A: Italian firms have the highest costs for electricity and the lowest overall cost for staff in the study

Panel B: Modeling—Smartphone services. *Sources* Juniper Research, AT Kearney, HSBC, IDC, Corporate reports (2011)

Panel C: Modeling—Aerospace. The profile of the Italian aerospace industry is similar to the UK with a strong impact from SMEs, but the sector is relatively smaller compared to the overall economy

5.5   Findings of Cloud Studies

The analysis of these cloud studies demonstrates the high degree of variability among countries based on the following three key factors. One is the fact that the choice of location of cloud services is strongly affected by energy prices and the propensity to use offshore facilities has both direct and indirect employment effects. Just as with the stimulus studies where the capacity of the domestic economy to meet forecast demand without great reliance on imports affects employment rates, the siting of data centers especially makes a difference on short-term employment for associated construction, equipping, and enhancing infrastructure.

A second finding is the very large discrepancy between sectors. This was anticipated and affected our choice of the mix of mature manufacturing, new services and infrastructure sectors, but even so the extent is such that it makes it all but senseless to generalize about the whole of an economy across many sectors. The factors that we took into consideration about growth rates based on analyses (some informed by surveys) of replacement cycles of service contracts as well as goods, and of redeployment as opposed to decrease in staffing, have direct consequences that differ by sector for employment trends.

Finally, we see that the direct positive employment effects are not dramatic except for those businesses that were founded on or later became dependent on cloud services. While all sectors pushed short-term employment through pressure to construct data centers, the longer term effects are easy to exaggerate. While we can show that they are in every case studied positive, and so contradict pessimistic speculations about the negative employment effects of new productivity-enhancing technologies, the scale is modest.

There is much literature already debunking the exaggerated claims that there is extensive pent-up demand for skilled labor,[12] nevertheless there remains much policy pressure to prioritize STEM education.[13] Especially considering the propensity to redeploy skilled staff, and a modest trend toward accepting that some more in-house training is worth spending on, there is some danger of accepting such exaggerations without taking into account the real capacity of these sectors to absorb available labor.

5.6   Digital London

The fourth study on employment effects started with a geographical question about the difference between London's growing rate of employment in technology and information jobs and trends in New York City and the San

Francisco/Silicon Valley regions. In contrast to the long-standing presence of a large and dynamic labor force in Northern California, London has grown dramatically only in recent years. In particular, growth in technology and information jobs constituted the major source of employment recovery following the financial crisis of 2008 and it changed the character of central London in important ways.

In the preceding studies of New York and San Francisco, my coauthor, Michael Mandel,[14] used a new method of analyzing job advertising data, along with census and industry sources, to locate jobs in two categories of jobs. The first of these we refer to as the "narrow" definition of technology and information jobs and identifies technicians and those doing work directly engaged in new technology applications to the business. The second, "broad", definition takes into account jobs associated with the new businesses that might not be directly technical but include design and marketing, support tasks and other jobs that would not exist but for the presence of the new technology workers.

We also considered the role of municipal and regional policies to take into account effects that might have come from direct subsidies or other market-distorting factors. In repositioning the methodology from the United States we needed to take into account differences in practice, such as local patterns of job advertising (we noted that advertisements tend to be listed considerably longer for the same jobs in the UK than in the United States).

For this study, the key starting point was to determine a workable definition of the tech/info sector and to do so for the UK in a way that would be sufficiently close to the US employment. The narrow definition is limited to one standard industrial category. For the UK, we included the Census Office's information and communications sector (SIC J): Which is constituted of publishing, software publishing (including games); motion picture, video, sound recording, and television program production; radio and television broadcasting; telecommunications (wired and wireless); computer programming and software development; computer consultancy; data processing and hosting; web portals; news agencies and other information service activities.[15]

The analysis of employment using the narrow definition is compared with the definition of the "expanded" info/tech sector. For the UK, this adds standard industrial code "M" (professional, scientific, and technical activities), in comparison with the addition for the US comparisons with NAICS 54 and 55 (professional, scientific, and technical services; management of companies and enterprises) (Table 5.11).

Table 5.11 What's driving London's growth?

	Numbers of jobs, 2013 (thousands)	Numbers of new jobs created, 2009–2013 (thousands)	Shares of total new jobs in London 2009–2013
Tech/info	382	39	8
Expended tech/info	1.088	143	30
Other industries	4.228	330	70

Note Expanded tech info includes tech/info, as noted in appendix
Data Office for National Statistics, South Mountain Economics LLC

When we compare the narrow with the expanded growth, we can see that in the four years following the 2008 financial crisis, fully 30% of all new jobs created in London are accounted for by tech/info (expanded) employment. Given the diversity of the London economy, this is highly disproportionate and reason to claim that the sector was heavily responsible for growth, especially in skilled and highly paid jobs.

However, not only is growth impressive for London, taken together, London, New York City, and San Francisco accounted for 41% of job growth in the combined US–UK tech–info sector 2009–2013. In that sense, London mirrors New York and San Francisco in that the growing tech/info sectors were a lifeline during the economic downturn from 2009. From that year, the tech/info sector grew 11% to 382,000 workers, a rate that was three times that of the previous four years. When taken together with kindred jobs, the "expanded" tech/info sector increased by 15% versus 8% for the rest of the London economy.

These years of growth created London as a major hub for big data with approximately 54,000 specialist workers within 25 miles of London.[16]

On a slightly smaller scale, but constituting the largest concentration in the world, fintech jobs employed approximately 44,000 workers within 25 miles of London, just slightly surpassing New York City, where 43,000 were employed in fintech. The significance of the link with the financial sector is particularly clear when seen in contrast to the San Francisco-Silicon Valley area, where there are 11,000 fintech workers.

The research not only reveals details about the scale and location of employment, it also holds implications, especially through comparative analysis, about why it has grown as it has, and how to keep it. There are, of course, several reasons for London's tech/info boom, and they resemble the reasons for the recent growth spurts in San Francisco and New York City. In part, this is a sort of organic growth with which tech companies share talent, ideas, products, and a labor market of well-educated workers. There are many talented programmers and designers, and many of them have been long employed in the local media firms and large, older tech companies. Most significantly, the customers of these goods and services are local in very long established sectors including publishing, the arts, advertising, education, and the public sector. There are also complementary industries, such as finance, media, and fashion that contribute to the growth of interlinked markets.

The presence of a reasonably well-installed infrastructure was also important for London, where the East End and Silicon Roundabout had appealing office space and good transportation and broadband connections. The local government contributed, as in the case of New York City, not through direct subsidies of great size, but through low-cost catalysts including concerted efforts to build up the image of London as a vibrant hub. There was a conscious effort to create a tech community through the formation of the Mayor's Tech City UK and related promotional and educational activities. These practices catalyzed a virtuous circle: Policy helped create clear guideposts for attracting start-ups, which generated more excitement, energy, and growth. This form of boosterism from municipal government was common to both London and New York, where Mayor Michael Bloomberg set a style similar to that adapted by Mayor Boris Johnson to put tech entrepreneurs at ease (Figs. 5.2, 5.3).

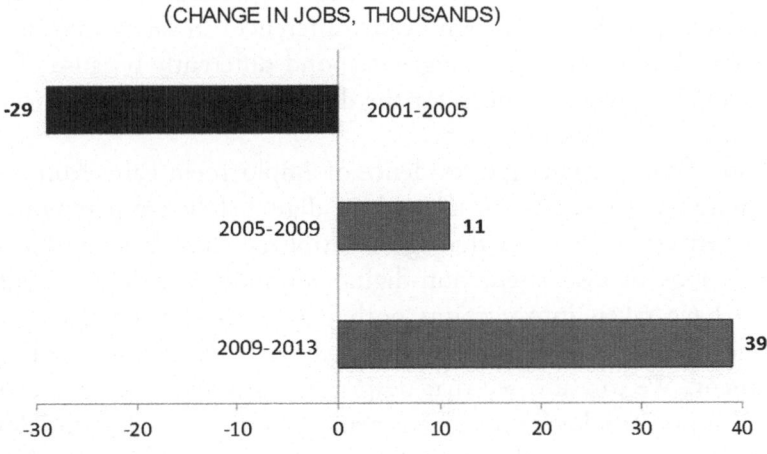

Fig. 5.2 London: Accelerating tech/info growth

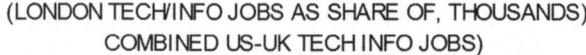

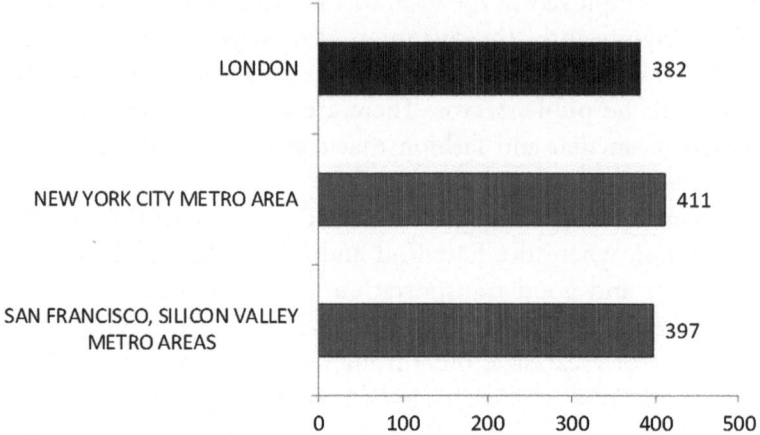

° Tech/Info employment as defined in appendix. Includes self-employed

Data: Office for National Statistics. Bureau of Labor statistics, South Mountain Economics LLC

Fig. 5.3   Global digital city: London gains ground

5.7   Synthesis

At the beginning, I raised four questions that these studies shed light on. (1) Given known investment intentions, how can an accelerated timescale boost sustainable employment? (2) To what extent do domestic capabilities along with sourcing policies make significant differences in labor trends? (3) To what extent do different technologies respond differently because of mix of job type. (4) Finally, we account for the distribution of effects between large versus small and medium size firms.

All four of these studies give evidence of employment gains from the digital economy, and not only in jobs where digital skills are paramount. The knock-on effects, as assessed through multipliers, show how unskilled jobs as well as jobs in associated, non-digital economy positions are affected. Since we have taken into account both job losses and job gains, we can have a fairly good sense that productivity gains are not going to reduce total employment. We can also see that despite the promise of cloud services to reduce in-house employment in data management, so long as the domestic economy has some capability to contribute to the provision of cloud services, total employment will not be reduced. The gains for cloud services, however, are very small and contingent on the domestic economic climate for the key inputs into running data centers and associated services: low energy costs and capable manpower.

The study of tech/info employment in London sheds light on two different trends. The first is that factors accelerating the growth of the digital economy are highly concentrated geographically and rest on the interrelationships among well-entrenched sectors (government, education, health care, media, retail, entertainment, and especially financial services) and the particular technological tools and trends currently driving growth (e.g., big data and fintech). The study also shows the significance of local factors such as municipal policies related to infrastructure and boosterism and other forms of "nudge" inducements.

These studies are significant for our understanding of the digital economy because they show how the focus on the character of the technology can make a difference to our analysis of the labor market trends. Not only do different areas of technology require different skills, they are differently entrenched into the broader economy in their own ways. Technologies in mature manufacturing industries, such as aerospace and automobile manufacturing, change more slowly and employment growth has smaller knock-on effects than those in new sectors, such as smartphone services. New technologies of cloud computing will reduce and redeploy in-house information management workers but the total effects, even when taking a conservative view of the productivity gains, is unlikely to reduce total employment of technically skilled people.

The policy implications of these studies are extensive. Contrary to both optimistic and pessimistic positions, we see a modest growth in most areas of digital economy employment. While it is not nearly large enough to solve the persistent unemployment problems of some economies, such as Italy and Turkey, it is significant in local areas, as can be seen by the dramatic boost to London's labor market brought on by the rise of "digital London". The fact that employment grows as we have described also raises many questions about the common, often shrill, warnings about massive skills shortages in the labor market. Where we have used salary data, we have not seen significant wage pressure in areas where we might have expected to note the effects of job-seekers applying strong bargaining power. Employers may claim that they cannot hire the people that they want, but they grow in any case, perhaps reluctantly resorting to investing in employee skills acquisition along the way.

The studies show some surprising results. Trade data and industry structure evidence indicate that the UK economy is more capable to meet the expectations of digital economy requirements in areas such as smart metering and intelligent transport systems infrastructure than might have been expected. Their reliance on imports may be less than sometimes assumed, and so fewer of the jobs generated by such spending will be created abroad. Another surprise was the strong effect that policies regarding restricting the export of data and on energy pricing have on the ability of cloud services to generate domestic employment. Especially given that, in theory, cloud services are highly portable, the ability to avoid exporting all new employment opportunities rests on the local economy's ability to achieve growth in those sectors that benefit from the new digital services as well as the providers of those services who rely upon data centers and skilled staff.

Notes

1. We have a right to be particularly cynical about employers' claims that the labor market is massively devoid of skilled workers because they have an interest in ensuring that in the long run, skills shortages cannot be used to create wage squeezes that drive up their salary bills. It is also apparent that the economy usually makes due with the labor force it has and still leaves some capable skilled workers unemployed or underemployed.

2. See the debate as expressed by Salzman, Hal, Daniel Kuehn, and B. Lindsay Lowell (2013), "Guestworkers in the High-Skill U.S. Labor Market, An Analysis of Supply, Employment, and Wage Trends," Economic Policy Institute, http://www.epi.org/files/2013/bp359-guestworkers-high-skill-labor-market-analysis.pdf; see also: Lazonick, William (2009), Sustainable Prosperity in the New Economy? Business Organization and High-Tech Employment in the United States. Kalamazoo, MI: Upjohn Institute of Employment Research.

3. Michael Mandel of the Progressive Policy Institute, who counts himself an optimist, is one of the few who use careful methodology to assess the employment effects of the digital economy to show how and where job creation can be found. His focus on the growth of fulfillment centers shows how low paid retail jobs has been multiplied by higher paid e-commerce related workers: http://www.progressivepolicy.org/blog/evolution-not-revolution-retail-apocalypse.

4. "The UK's Digital Road to Recovery" following the parallel US "stimulus" study by ITIF (supported by IBM funding 2009).
   Modeling the Cloud; employment effects in two exemplary sectors in the US, UK, Germany & Italy," which adjusted the ITIF model for a bottom-up analysis of labor changes from investments in specific technologies (supported by Microsoft funding 2012).

"Modeling the Cloud for Turkey," which applied to specific national conditions (Microsoft funding 2013).
"London: Digital City on the Rise" w/Michael Mandel paralleling his NYC and San Francisco studies (Bloomberg Philanthropies funding 2014).

5. Etro, F. (2009), "The Economics of Cloud Computing on Business Creation, Employment and Output in Europe," *Review of Business and Economics* 2: 179–208.

6. It has also been used, more controversially, to inflate the economic effects of certain policies. An example of that which is directly relevant to this work is the report by NESTA (2010) which used distinctly nonconservative multipliers to imply that employment effects of broadband investment can be huge—on the order of ten times my estimates. Kapstein has used this approach most convincingly in a series of reports on the economic impact of foreign direct investment, for example, Kapstein, E. (2008), "Measuring Unilever's Economic Footprint: The Case of South Africa," Unilever.

7. In a related study, we surveyed chief information officers or related executives in large enterprises in seven countries about, among other things, their intentions to redeploy or reduce staff as a consequence of new productivity-enhancing methods including cloud computing services, outsourcing opportunities, etc., Liebenau and Karrberg (2008), "Enterprise Efficiency in the Use of ICT in China, France, Germany, Great Britain, India, Japan & the USA," LSE Enterprise (with funding from Dell), https://www.researchgate.net/publication/277188210_Enterprise_efficiency_in_the_use_of_ICT_in_China_France_Germany_Great_Britain_India_Japan_the_USA_first_interim_report_on_LSE-Dell_research.

8. We assume that the short-term effects on the proportion of small versus large firms stay constant, i.e., that there is no special boost for SMEs.

9. Independent research conducted with ITIF using IBM funding and mainly UK statistical sources, ILU and OECD employment data, industry and company-level data on tasks, pay, and firm capabilities.

10. J.M. Keynes is said to have pointed out, probably factiously, that the most straightforward way to boost employment is to dig holes and fill them in again, an observation that perhaps influenced some "stimulus spending" advisors to focus especially on infrastructure construction or maintenance projects at the expense of less immediately efficacious job creation in new industry development activities.

11. Both studies were independent research supported by Microsoft funding and based on national statistical sources, ILU, OECD employment data, industry and company-level data on tasks, pay, firm capabilities, etc.

12. See for example, Lowell, B. Lindsay, and Hal Salzman (2007), "Into the Eye of the Storm: Assessing the Evidence on Science and Engineering Education, Quality, and Workforce Demand," Research Report, Urban Institute, http://www.urban.org/publications/411562.html.

13. Atkinson, Robert D., and Merrilea Mayo (2010), "Refueling the U.S. Innovation Economy: Fresh Approaches to Science, Technology, Engineering and Mathematics (STEM) Education," Washington, DC, ITIF. However, consider the oversupply of STEM skilled youth in some countries, notably South Korea, this concern is not fanciful.

14. Mandel, Michael (2014), "San Francisco and the Tech-Info Boom: Making the Transition to a Balanced and Growing Economy," South Mountain Economics. https://southmountaineconomics.files.wordpress.com/2014/05/sf-techinfo-study.pdf; Mandel, Michael (2013), "Building a Digital City: The Growth and Impact of New York City's Tech/Information Sector," South Mountain Economics. https://southmountaineconomics.files.wordpress.com/2013/09/building-a-digital-city1.pdf.

15. While there is a large degree of overlap with the comparable list for the United States under the category "information" (NAICS 51), that list consists of: publishing; software publishing; motion picture and sound recording; radio, television, and cable broadcasting; telecommunications (wired and wireless); data processing and hosting; internet publishing and broadcasting and web search portals, news syndicates, and other information services. This is combined with the category "computer systems design and related services" (NAICS 5415): custom computer programming services; computer systems design services; computer facilities management services.

16. This compares with New York City which has slightly more big data workers (approx. 57,000), but both are significantly behind the San Francisco-Silicon Valley region where 98,000 work in big data. Presumably, they are engaged in the large social media, advertising and online commerce activities concentrated in California.

Bibliography

Atkinson, Robert D., Daniel Castro, and Stephen Ezell. 2009. *The Digital Road to Recovery: A Stimulus Plan to Create Jobs, Boost Productivity and Revitalize America*. Washington, DC: ITIF.

Atkinson, Robert D., and Merrilea Mayo. 2010. *Refueling the U.S. Innovation Economy: Fresh Approaches to Science, Technology, Engineering and Mathematics (STEM) Education*. Washington, DC: ITIF.

Brynjolfsson, Eric, and Andrew McAfee. 2011. *Race Against the Machine*. Lexington, MA: Digital Frontier Press.

Etro, Frederico. 2009. "The Economics of Cloud Computing on Business Creation, Employment and Output in Europe." *Review of Business and Economics* 2: 179–208.

Kapstein, Ethan. 2008. *Measuring Unilever's Economic Footprint: The Case of South Africa*. London: Unilever.

Lazonick, William. 2009. *Sustainable Prosperity in the New Economy? Business Organization and High-Tech Employment in the United States*. Kalamazoo, MI: Upjohn Institute of Employment Research.

Liebenau, Jonathan, and Patrik Karrberg. 2008. "Enterprise Efficiency in the Use of ICT in China, France, Germany, Great Britain, India, Japan & the USA." LSE Enterprise (with Funding from Dell). https://www.researchgate.net/publication/277188210_Enterprise_efficiency_in_the_use_of_ICT_in_China_France_Germany_Great_Britain_India_Japan_the_USA_first_interim_report_on_LSE-Dell_research.

Liebenau, Jonathan, Robert Atkinson, Patrik Karrberg, and Daniel Castro. 2009. *The UK's Digital Road to Recovery*. Washington, DC: ITIF.

Liebenau, Jonathan, and Patrik Karrberg. 2012. *Modeling the Cloud; Employment Effects in Two Exemplary Sectors in the US, UK, Germany & Italy*. London: LSE Enterprise.

Liebenau, Jonathan, and Patrik Karrberg. 2013. *Modeling the Cloud for Turkey*. London: LSE Enterprise.

Mandel, Michael. 2013. "Building a Digital City: The Growth and Impact of New York City's Tech/Information Sector." Washington, DC: South Mountain Economics. https://southmountaineconomics.files.wordpress.com/2013/09/building-a-digital-city1.pdf.

Mandel, Michael. 2014. "San Francisco and the Tech-Info Boom: Making the Transition to a Balanced and Growing Economy." Washington, DC: South Mountain Economics. https://southmountaineconomics.files.wordpress.com/2014/05/sf-techinfo-study.pdf.

Mandel, Michael, and Jonathan Liebenau. 2014. "London: Digital City on the Rise." Washington, DC: South Mountain and Bloomberg Philanthropies.

Michaels, Guy, Ashwini Natrajm, and John Van Reenen. 2014. "Has ICT Polarized Skill Demand? Evidence from Eleven Countries over 25 Years." *The Review of Economics and Statistics* 96 (1): 60–77.

Salzman, Hal, Daniel Kuehn, and B. Lindsay Lowell. 2013. "Guestworkers in the High-Skill U.S. Labor Market, an Analysis of Supply, Employment, and Wage Trends." Washington, DC: Economic Policy Institute. http://www.epi.org/files/2013/bp359-guestworkers-high-skill-labor-market-analysis.pdf.

6

The Impact of the Broadband Internet on Employment

Raul L. Katz

6.1 Introduction

The topic of the impact of the broadband Internet on employment has been present in the public policy arena quite regularly in the past years. Unfortunately, such an important debate has been approached with little formalization of the research of impact and understanding of the evidence. Before even tackling the prescriptive side of the policy debate, researchers appear to be aligned in two camps: The Internet contributes to the creation of jobs, and, on the opposite side, the Internet is the source of job destruction. Unfortunately, in many cases, research is being conducted hypothetically (e.g., what kind of jobs are susceptible to be eliminated as a result of digitization?[1]) without looking at the evidence of what has occurred since the Internet has become widely adopted by consumers and enterprises. This chapter summarizes the results of investigations conducted by this author and other researchers with regard to impact of broadband on employment.[2]

This chapter argues that, based on the evidence, the response to the question of impact of broadband on employment is: it depends. In fact, it will be shown that broadband contributes to the creation of jobs in certain industries and geographies, while also being a key factor in capital-labor substitution under certain conditions. As is always the case, this kind of answer does

R. L. Katz (✉)
Columbia Institute for Tele-Information,
Columbia Business School, New York, USA

© The Author(s) 2018
L. Pupillo et al. (eds.), *Digitized Labor*, https://doi.org/10.1007/978-3-319-78420-5_6

not satisfy pundits or ideologues. However, in the end, if policy makers are oriented toward making good decisions, they need to have a solid, unbiased understanding of the evidence.

6.2   What Does Our Research Tell Us About Employment Effects of Broadband?

Broadband can have a positive effect on job creation under certain circumstances. To begin with, broadband deployment programs tend to create jobs under attractive multipliers on a short-term basis. We tend to call this the "construction effect", which has been put in place with good results as a countercyclical measure. Second, deployment of broadband in emerging countries enables these nations to attract employment (especially low paid business process outsourcing jobs) from industrialized economies. It could be argued, however, that this represents a zero-sum game since an emerging country gain is an industrialized nation loss.

A third job creation effect of broadband has been identified in advanced economies. It refers to the emergence of broadband-enabled businesses that were previously nonexistent, such as Internet search and advertising and electronic commerce. Finally, even within industries that predate the Internet, broadband has generated spill-overs with regards to job creation. To clarify, broadband enables business to redeploy functions in order to achieve better economics: This could lead to the creation of employment in certain regions in order to benefit from the availability of wider labor pools or lower factor costs. We acknowledge that this is again a case of zero-sum (a job loss in a metropolitan area represents a gain in a suburban or rural zone). However, we have detected a fifth effect, which we could refer to as the market reach. In this case, firms can rely on broadband to deploy distribution channels in otherwise unserved remote geographies. We do not refer here to electronic commerce channels since these firms do not need a physical presence to reach remote areas but consider industries that still require some brick and mortar (and consequently employees) to deliver a certain service. An example of this effect can be found in the health-care sector: broadband represents an enabling technology allowing hospitals to deploy "satellite" clinics in remote areas charged with delivery health-care services while benefitting from accessing technical and clerical support from a central facility. We have so far outlined job creation effects related to broadband technology. Let's now move job destruction cases.

First and foremost, we recognize the other side of the zero-sum game of low paid jobs in industrialized countries being outsourced to emerging nations to exploit lower factor costs. Second, we have found both in industrialized and emerging countries strong capital/labor substitution effects in labor-intensive sectors. A case in point is the tourism industry, where broadband has contributed to reduce employment across the board. Third, one should not underestimate job losses in rural geographies resulting from broadband adoption. While job losses driven by productivity enhancements in metropolitan areas can be compensated by innovation-led new business models or natural expansion, rural settings lack this mechanism.

As it can be seen, the answer to the job creation/destruction question is: It depends on the sector, geography, and overall stage of economic development. We will now move to detail the empirical evidence in support of each of these effects.

6.3   The "Construction" Effect

There are three types of job creation effects resulting from broadband network construction. The first, most straightforward one comprises employment growth generated in the course of deployment of network infrastructure. Jobs in this area typically entail telecommunications technicians, construction workers and civil and radio frequency engineers. The second job creation effect captures indirect jobs triggered by network construction. It entails employment generated by indirect spending or businesses buying and selling to each other in support of direct network rollout. Jobs created through this effect include metal products and electrical equipment workers, as well as professional services. Finally, the third job creation effect resulting from network deployment comprises jobs induced by household spending based on the income earned from the direct and indirect effects. In this case, we are referring to employment in consumer durables, retail trade, and consumer services.

Four national studies have estimated the impact of network construction on job creation: Crandall et al. (2003), Atkinson et al. (2009), Liebenau et al. (2009), and Katz et al. (2008). They all relied on input–output matrices and assumed a given amount of capital investment: USD 63 billion (needed to reach ubiquitous broadband service in the United States) for Crandall et al. (2003), CHF 13 billion for Katz et al. (2008) (to build a national multi-fiber network for Switzerland), USD 10 billion for Atkinson et al. (2009) (as a US broadband stimulus) and USD 7.5 billion for Liebenau et al. (2009) (needed to complete broadband deployment in the United Kingdom) (see Table 6.1).

Table 6.1 Broadband impact on job creation (*Sources* Katz, R., and S. Suter (2009), Estimating the Economic Impact of the US Broadband Stimulus Plan, Columbia Institute for Tele-Information Working Paper; Katz, R., P. Zenhäusern, S. Suter, P. Mahler, and S. Vaterlaus (2008), *Economic Modeling of the Investment in FTTH in Switzerland*, unpublished report; Libenau, J., and R. Atkinson (2009), *The UK's Digital Road to Recovery.* LSE and ITIF; Australian government. Katz, R., Vaterlaus, P. Zenhäusern, S. Suter, and P. Mahler (2009), *The Impact of Broadband on Jobs and the German Economy*, Columbia Institute for Tele-Information Working Paper)

Country	Authors—institution[a]	Objective	Results
United States	Crandall et al. (2003)—Brookings Institution	Estimate the employment impact of broadband deployment aimed at increasing household adoption from 60 to 95%, requiring an investment of USD 63.6 billion	• Creation of 140,000 jobs per year over ten years • Total jobs: 1.2 million (including 546,000 for construction and 665,000 indirect)
	Atkinson et al. (2009)—ITIF	Estimate the impact of a USD 10 billion investment in broadband deployment	• Total jobs: 180,000 jobs-year (including 64,000 direct and 116,000 indirect and induced)
Switzerland	Katz et al. (2008)—CITI	Estimate the impact of deploying a national broadband network requiring an investment of CHF 13 billion	• Total jobs: 114,000 over four years (including 83,000 direct and 31,000 indirect)
United Kingdom	Liebenau et al. (2009)—LSE	Estimate the impact of investing USD 7.5 billion to achieve the target of the *"Digital Britain"* Plan	• Total jobs: 211,000 jobs-year (including 76,500 direct and 134,500 indirect and induced)
Australia	Australian government	Estimate the impact of investing USD 31,340 million to deploy high speed broadband	• Total jobs: ~ 200,000 jobs-year

[a]*ITIF* Information Technology and Innovation Foundation, *CITI* Columbia Institute for Tele-Information, *LSE* London School of Economics

Since these studies were triggered by the consideration of countercyclical plans devised to face the 2008 economic crisis, they tended to focus primarily on gauging the ability of broadband to create jobs. All studies calculated multipliers, which measure the total employment change throughout the economy resulting from the deployment of a broadband network. Multipliers are of two types: Type I multipliers measure the direct and indirect effects (direct plus indirect divided by the direct effect), while type II multipliers measure type I effects plus induced effects (direct plus indirect plus induced divided by the direct effect). Cognizant that multipliers from one geographic region cannot be applied to another, it is useful to observe the summary results for the multipliers of the four input–output studies (see Table 6.2).

According to the sector interrelationships as depicted above, European economies would appear to have lower indirect effects than the United States. Furthermore, the disaggregation of effects also indicates that a relatively important job creation induced effect occurs as a result of household spending based on the income earned from direct and indirect jobs.[3]

While input–output tables are a reliable tool for predicting investment impact, two words of caution need to be given. First, input–output tables are static models reflecting the interrelationship between economic sectors at a certain point in time. Since those interactions may change over time, the matrices could lead us to overestimate or underestimate the impact of

Table 6.2 Employment multiplier effects of studies relying on input–output analysis (*Source* Katz, R., and S. Suter (2009), Estimating the Economic Impact of the US Broadband Stimulus Plan, Columbia Institute for Tele-Information Working Paper; Katz, R., P. Zenhäusern, S. Suter, P. Mahler, and S. Vaterlaus (2008), *Economic Modeling of the Investment in FTTH in Switzerland*, unpublished report; Liebenau, J., Atkinson, R. (2009). *The UK's Digital Road to Recovery*. LSE and ITIF; Australian government. Katz, R., S. Vaterlaus, P. Zenhäusern, S. Suter, and P. Mahler (2009), *The Impact of Broadband on Jobs and the German Economy*; Columbia Institute for Tele-Information Working Paper)

Country	Studies	Type I	Type II
United States	Crandall et al. (2003)	N.A.	2.17
	Atkinson et al. (2009)	N.A.	3.60
	Katz et al. (2009)	1.83	3.42
Switzerland	Katz et al. (2008)	1.38	N.A.
United Kingdom	Liebenau et al. (2009)	N.A.	2.76
Germany	Katz et al. (2010)	1.45	1.92

N.A. Not Available
Note Crandall et al. (2003) and Atkinson et al. (2009) do not differentiate between indirect and induced effects, therefore we cannot calculate Type I multipliers; Katz et al. (2008) did not calculate Type II multiplier because induced effects were not estimated

network construction. For example, if the electronic equipment industry is outsourcing jobs overseas at a fast pace, the employment impact of broadband deployment will diminish over time and part of the countercyclical investment will "leak" overseas. Second, it is critical to break down employment effects at the three levels estimated by the input–output table in order to gauge the true direct impact of broadband deployment. Having said that, all these effects have been codified and therefore, with the caveat of the static nature of input–output tables, we believe that the results were quite reliable.

6.4   Job Creation Resulting from Broadband Spillovers

Beyond the employment and output impact of network construction, researchers have also studied the impact of network externalities on employment variously categorized as "innovation", or "network effects".[4] The study of network externalities resulting from broadband penetration has led to the identification of numerous effects:

- New and innovative applications and services, such as telemedicine, Internet search, e-commerce, online education, and social networking[5]
- New forms of commerce and financial intermediation[6]
- Mass customization of products[7]
- Marketing of excess inventories and optimization of supply chains[8]
- Business revenue growth[9]
- Growth in service industries[10]

The evidence regarding broadband employment externalities also appears to be quite conclusive (see Table 6.3).

The spillover impact of broadband on employment creation appears to be positive. However, as the evidence indicates, the impact on employment growth varies widely, from 0.2 to 5.32% for every increase in 1% of penetration. There are several explanations for this variance. As Crandall indicated, the overestimation of employment creation in his study is due to employment and migratory trends, which existed at the time and biased the sample data. In the case of Gillett et al. (2006), researchers should be careful about analyzing local effects because zip codes are small enough areas that cross-zip code commuting might throw off estimates on the effect of broadband. For example, increased wages from broadband adoption in one zip code would probably raise rent levels in neighboring zip codes prompting some migration effects. Finally, the wide range of effects in the case of Shideler et al. (2007)

Table 6.3   Research results of broadband impact on employment in the United States (*Source* Author)

Study	Data	Effect
Crandall et al. (2007)	48 US states for the period 2003–2005	For every 1% point increase in broadband penetration in a state, employment is projected to increase by 0.2–0.3% per year "assuming the economy is not already at 'full employment'"
Thompson and Garbacz (2008)	46 US states during the period 2001–2005	Positive employment generation effect varying by industry
Gillett et al. (2006)	US zip codes for the period 1999–2002	Broadband availability increases employment by 1.5%
Shideler et al. (2007)	Disaggregated county data for state of Kentucky for 2003–2004	An increase in broadband penetration of 1% contributes to total employment growth ranging from 0.14 to 5.32% depending on the industry

is explained by the divergent effects among industry sectors. We will explore this particular effect in turn.

6.5   Differential Employment Impact by Industry Sector

As with output, the spillover employment effects of broadband are not uniform across sectors. Two studies have identified differential levels of impact. According to Crandall et al. (2007), the job creation impact of broadband tends to be concentrated in service industries (e.g., financial services, education, health care, etc.) although the authors also identified a positive effect in manufacturing (see Table 6.4).

In another study, Shideler et al. (2007) found that, for the state of Kentucky, county employment was positively related to broadband adoption in the following sectors (see Table 6.5).

The only sector where a negative relationship was found with the deployment of broadband (0.34–39.68%) was the accommodations and food services industry. This may result from a particularly strong capital/labor

Table 6.4 Coefficient of broadband penetration in employment growth by sector (with significance at the 5% and 1% confidence level) (*Source* Crandall et al. 2007)

Sector	Employment 2005–4		Employment 2005–3	
	Coefficient	T-statistic	Coefficient	T-statistic
Manufacturing	0.371	2.46	0.789	2.59
Educational services	2.741	2.73	4.054	3.25
Health care	3.369	2.50	0.656	2.51
Accommodation and food services	0.284	2.12	N.A.	
Finance and insurance	N.A.		1.043	3.09

N.A. Statistically not significant

Table 6.5 Kentucky: Differential impact of broadband by industry sector (*Source* Shideler et al. 2007)

Sector	95% Confidence interval (%)
Aggregate	0.14–5.32
Construction	0.62–21.76
Information	25.27–87.07
Administrative	23.74–84.56

substitution process, whereby productivity gains from broadband adoption yields reduced employment. Crandall et al. (2007) also found a negative relationship for the Arts, Entertainment & Recreation sector, although it was not statistically significant. Similarly, Thompson and Garbacz (2008) concluded that, for certain industries, "there may be a substitution effect between broadband and employment".[11] It should therefore be considered that the productivity impact of broadband can cause capital-labor substitution and may result in a net reduction in employment.

In summary, research pinpoints different employment effects by industry sector. Broadband may simultaneously cause labor creation triggered by innovation in services and a productivity effect in labor-intensive sectors. In light of these effects, given that the sector composition varies by regional economies, the deployment of broadband should not have a uniform impact across a national territory.

6.6   Differential Employment Impact by Region

In two studies conducted by this author, it was found that, as expected, employment impact of broadband technology varies by region of a country.

In study conducted with German data by Länder (counties) split between high broadband penetrated and low broadband penetrated counties, it was found that employment impact varied significantly by region (see Table 6.6).

In high broadband penetrated counties the short-term impact of the technology is very high both on GDP and employment, but it declines over time. This "supply shock" is believed to occur because the economy can immediately utilize the new deployed technology. Furthermore, the fact that employment and GDP grow in parallel indicates that broadband has a significant impact on innovation and business growth, thereby overcoming any employment reduction resulting from productivity effects.

On the other hand, in counties with low broadband penetration the impact on GDP of broadband penetration is lower than in high-penetrated areas in the short term, but "catches up" to comparable levels over time. The impact of broadband on employment is slightly negative in the initial years.

Table 6.6 Germany: Comparative effects between high broadband and low broadband counties (*Source* Katz et al. 2010)

	Total	Low penetration	High penetration
Growth of GDP[a]			
GDP per Capita 2000 (*1,000,000)	0.0261	0.0627	0.0185
	(0.041)	(0.121)	(0.050)
Population growth (2000–2006)	0.6318***	0.5311***	0.7731***
	(0.075)	(0.102)	(0.116)
Broadband penetration growth (2002–2003)	0.0255***	0.0238***	0.0256***
	(0.002)	(0.005)	(0.003)
R^2 adjusted	0.6317	0.6321	0.6305
Number of observations	424	210	214
Growth of Employment[b]			
GDP per Capita 2000 (*1,000,000)	0.0362*	−0.0066	0.0030
	(0.024)	(0.072)	(0.029)
Population growth (2000–2006)	1.0481***	1.1265***	0.9072***
	(0.044)	(0.061)	(0.066)
Broadband penetration growth (2002–2003)	0.0020*	0.0027	0.0061***
	(0.001)	(0.003)	(0.002)
R^2 adjusted	0.6065	0.6597	0.5557
Number of observations	424	210	214

[a]Dependent Variable: Growth of GDP between 2003 and 2006.
G_GDP (03–06) = $\beta 1 \times$ GDP_Capita_2000 + $\beta 2 \times$ G_POP (00–06) + $\beta 3 \times$ G_BBPEN (02–03).
[b]Dependent Variable: Growth of Employment between 2003 and 2006.
G_EMP (03–06) = $\beta 1 \times$ GDP_Capita_2000 + $\beta 2 \times$ G_POP (00–06) + $\beta 3 \times$ G_BBPEN (02–03).
***, ** and * indicate a significance level of 5%, 10% and 15%. Standard errors in parenthesis

This indicates that the impact of broadband in low penetration areas is more complex than in the high penetration areas. The increase in broadband penetration in low penetrated areas takes longer to result in economic growth because these economies require a longer period of time to develop and fully utilize the technology. However, after three years the level of impact of broadband in low penetrated regions is as high as in the more developed areas. Negative initial employment growth appears to indicate that the productivity increase resulting from the introduction of new technology is the most important effect to begin with. However, once the economy develops, the other network effects (innovation and value chain recomposition) start to play a more important role, resulting in job creation.[12] Therefore broadband deployment in low-penetrated areas will likely generate high stable economic growth ("catch up" effect) combined capital/labor substitution, which initially limits employment growth ("productivity" effect). Figure 6.1 presents in conceptual fashion a comparison of impact in both regions.

A similar differentiated effect was found by this author in a study of broadband impact in the state of Kentucky. Similarly to Kandilow and Renkow (2010) results regarding broadband loans in rural areas, it was found that the impact of broadband availability is dependent upon the area of deployment. Katz et al. (2012) have found that while though broadband availability impacts rural as well as metropolitan counties, the effect appears to be area-specific (see Table 6.7).

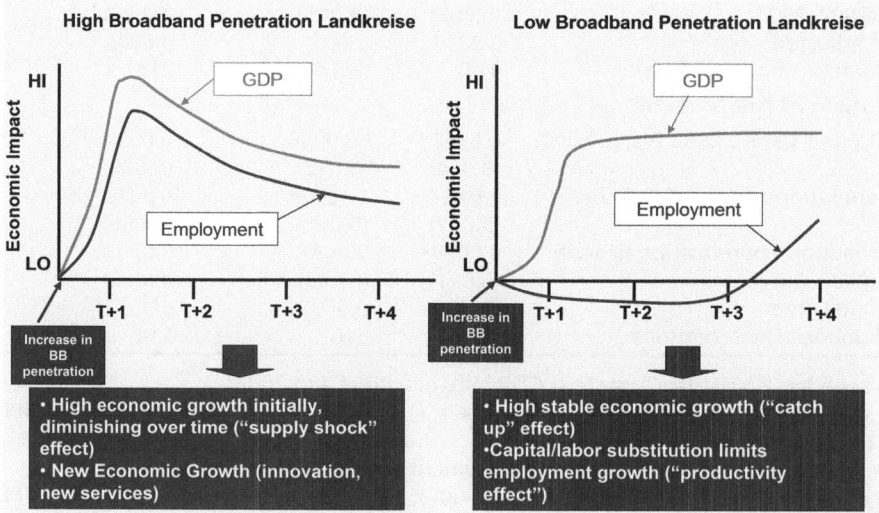

Fig. 6.1 Conceptual view of comparative broadband regional effects (Only effects up to t + 3 are estimated.) (*Source* Adapted from Katz et al. 2010)

Table 6.7 Kentucky: Impact of a 1% increase in broadband availability on employment and median income (*Source* Data compiled from Connect Kentucky databases, and ESRI Business Analyst Sourcebook for County demographics; Katz et al. 2012)

	Impact on median income	Impact on employment
Metropolitan counties	0.0968*	0.0301
Rural counties adjacent to metro counties	0.0704*	−0.1953*
Rural counties isolated from metro counties	0.0800*	

*Significant at the 1% level

The model results show that the impact of broadband on median income is statistically significant for each of the three types of counties. They also suggest that this impact is the highest for metro counties, followed by isolated rural counties, and lastly rural counties that are adjacent to metro counties. On the other hand, the impact on unemployment is only significant for rural counties.[13] This is a reasonable result in light of the merging of labor markets. In this context, it is to be expected that broadband will have the smallest impact on metro counties. These counties have the lion's share of establishments and employment opportunities so increasing the size of the labor market should have only marginal if any positive effects. However, broadband may extend labor markets to rural areas, for example, by enabling telecommuting. Of these rural counties, the primary beneficiaries are rural counties that are adjacent to metro areas because the labor force is more technologically skilled (in accordance with the industries that are present). We expect that isolated rural areas will also benefit, but at a lower rate.

Theoretically, we also expect that firms in the services industries can reap greater productivity gains from broadband (see below for the results on sector-specific broadband effects). Hence it is expected that metro counties, which account for the vast majority of such firms, will experience the largest impact on income. This indicates that the employment opportunities created by broadband in these areas are far more lucrative than the median job. Though the portion of the population that is technologically skilled in these areas may be small, it is likely that the incremental benefits of broadband for this population are quite high. However, it was not possible to identify a statistically significant result for metro counties.

The impact of broadband penetration was found to be statistically significant on the growth in employment in the financial services and insurance, wholesale trade, and health sectors (see Table 6.8).

The results of the sector impact models are quite illuminating in terms of determining which industries are most benefited by rural broadband.

Table 6.8 Kentucky: Impact of broadband penetration by 1% on industrial sector employment (*Source* Data compiled from US Census Bureau, Connect Kentucky databases, and ESRI Business Analyst Sourcebook for County demographics; Katz et al. 2012)

Industry sector	All counties	Rural counties
Financial services and insurance	0.678**	0.517***
Wholesale trade	0.846*	0.836*
Health services	0.126*	0.122**
Construction	Not significant	Not significant
Retail trade	Not significant	Not significant
Accommodation	Not significant	Not significant

*Significant at 1% level, **Significant at 5% level, ***Significant at 10% level

While effects are statistically significant in finance, wholesale trade, and health services, the impact is largest in the trade sector, reflecting the value of broadband as an enabler of relocation of warehouses and distribution centers to areas outside the metropolitan counties. Furthermore, while employment is also positively impacted by broadband in finance, its contribution diminishes in rural environments reflecting the difficulty of locating financial back offices in rural areas, primarily due to limits in labor pool availability. On the other hand, the decline in impact of health services for rural areas is not that important revealing both the existence of demand in rural areas and the value of broadband in enabling the redeployment of health facilities.

6.7 Conclusion

To conclude, the evidence regarding employment impact of broadband Internet underlines the danger of reaching uniform deterministic answers. Deployment of broadband networks has a short-term Keynesian effect, while spillover impact requires a much longer time frame to materialize. Moreover, externalities tend to vary substantially by geography and industrial sector. One can assume that similar conclusions could extend to the impact of job creation across the digital ecosystem. For example, research indicates that, direct job creation effect of digital platforms appears to be fairly limited,[14] while indirect employment created as a result of new firm creation is fairly large.[15]

This evidence points out to the need to develop public policies that promote positive effects in terms of job creation, while mitigating the negative ones. For example, nations undergoing important broadband network deployment efforts should consider implementing conventional rural development programs aimed at reducing "hollowing out" effects. In the case of emerging nations, it would be convenient to centralize digital policy development and implementation in order to control for potential job losses in certain sectors and geographies.

Notes

1. See Frey, C., and M. Osborne (2013), *The Future of Employment: How Susceptible Are Jobs to Computerization*. Oxford University Martin School.
2. The difference between the causal factor (broadband or Internet) is quite important. While broadband is the telecommunication technology required to access the web, it also contributes to the communication among individuals, enterprises, and government agencies.
3. It is assumed that induced effects should be counted since in 2008, no full employment conditions existed.
4. Atkinson et al. (2009).
5. Op. cit.
6. Op. cit.
7. Op. cit.
8. Op. cit.
9. Varian et al. (2002), Gillett et al. (2006).
10. Crandall et al. (2007).
11. This effect was also mentioned by Gillett et al. (2006).
12. This said, the available data sets do not enable us to test this last point at this time.
13. The models run for employment impact on rural-adjacent and rural-isolated yielded nonsignificant results.
14. By 2014, Google, Facebook, Skype, Twitter, LinkedIn, and Netflix accounted for a total headcount of 68,885 (*Source* Annual reports).
15. In Latin America alone, the video game industry created 120,000 jobs (*Source* Katz 2015).

References

Atkinson, R., D. Castro, and S.J. Ezell. 2009. *The Digital Road to Recovery: A Stimulus Plan to Create Jobs, Boost Productivity and Revitalize America*. Washington, DC: The Information Technology and Innovation Foundation.

Crandall, R., C. Jackson, and H. Singer. 2003. *The Effect of Ubiquitous Broadband Adoption on Investment, Jobs, and the U.S. Economy*. Washington, DC: Criterion Economics.

Crandall, R., W. Lehr, and R. Litan. 2007. "The Effects of Broadband Deployment on Output and Employment: A Cross-Sectional Analysis of U.S. Data." *Issues in Economic Policy*, 6.

Frey, C., and M. Osborne. 2013. *The Future of Employment: How Susceptible Are Jobs to Computerization?* Oxford, UK: Oxford University Martin School.

Gillett, S., W. Lehr, C. Osorio, and M.A. Sirbu. 2006. Measuring Broadband's Economic Impact. Technical Report 99-07-13829, National Technical Assistance, Training, Research, and Evaluation Project.

Kandilow, I., and M. Renkow. 2010. "Infrastructure Investment and Rural Economic Development: An Evaluation of USDA's Broadband Loan Program." *Growth and Change* 41 (2): 165–191.

Katz, R. 2015. *El ecosistema y la economía digital en América Latina*. Madrid: Ariel.

Katz, R.L., and S. Suter. 2009. Estimating the Economic Impact of the Broadband Stimulus Plan. Columbia Institute for Tele-Information Working Paper. http://www.elinoam.com/raulkatz/Dr_Raul_Katz_-_BB_Stimulus_Working_Paper.pdf.

Katz, R.L., P. Zenhäusern, and S. Suter. 2008. *An Evaluation of Socio-Economic Impact of a Fiber Network in Switzerland*. Polynomics and Telecom Advisory Services, LLC, New York.

Katz, R.L., S. Vaterlaus, P. Zenhäusern, and S. Suter. 2010. The Impact of Broadband on Jobs and the German Economy. *Intereconomics* 45 (1): 26–34.

Katz, R., J. Avila, and G. Meille. 2012. *Economic Impact of Wireless Broadband in Rural America*. Washington, DC: Competitive Carriers Association.

Liebenau, J., R.D. Atkinson, P. Kärrberg, D. Castro, and S.J. Ezell. 2009. *The UK's Digital Road to Recovery*. April 29. https://ssrn.com/abstract=1396687.

Shideler, D., N. Badasyan, and L. Taylor. 2007. "The Economic Impact of Broadband Deployment in Kentucky." Telecommunication Policy Research Conference, September 28–30, Washington, DC.

Thompson, H.G., and C. Garbacz. 2008. Broadband Impacts on State GDP: Direct and Indirect Impacts. Paper presented at the 17th Biennial Conference of the International Telecommunications Society (ITS), June 24–27, Montreal.

Varian, H., R. Litan, A. Elder, and J. Shutter. 2002. *The Net Impact Study: The Projected Economic Benefits of the Internet in the United States, United Kingdom, France and Germany*. https://www.cisco.com.

7

The Impact of the Internet on Employment and Income in the US Media and Entertainment Business

David Viviano

7.1 Introduction

The US Media and Entertainment sector is vibrant and dynamic. By a variety of metrics it has been growing steadily for decades and consistently represents a positive net export (Vogel 2014). SAG-AFTRA is a US labor union that represents the talent that creates content for this industry: actors, journalists, recording artists, and other performers in film, television, radio, and digital media. SAG-AFTRA's members include a wide diversity of media talent: the Oscar-winning movie stars of big-budget Hollywood films, Grammy-winning recording artists, the character actors that are recognizable from various television shows, stunt performers, and thousands of other media talent, most of whom don't have much name recognition.

SAG-AFTRA negotiates employment agreements with various industry employers on behalf of its members for terms of minimum compensation and working conditions across a wide array of media work. SAG-AFTRA also negotiates the terms for residuals payments when its members' work is re-used or exhibited in a different market. While SAG-AFTRA is a US-based labor union, it ensures that its members are protected by the terms of its contracts wherever in the world they may work. As the representative

D. Viviano (✉)
Screen Actors Guild-American Federation of Television and Radio Artists (SAG-AFTRA), Los Angeles, CA, USA

© The Author(s) 2018
L. Pupillo et al. (eds.), *Digitized Labor*, https://doi.org/10.1007/978-3-319-78420-5_7

for these workers, SAG-AFTRA has unique access to granular employment and earnings data for this work.

The Internet, digital technology and digital media have had myriad impacts on the SAG-AFTRA talent who work in the media sector. The following analysis attempts to explain how the Internet has impacted the demand for labor, the supply of labor, and changes in the net employment and income for this population.

7.2   Consumption of Media

Understanding changes in how media is being consumed is important for understanding what is driving changes in the demand for labor. One truly revolutionary change in the media business that started decades ago is the growing "shelf space" for content. Before Multichannel Video Programming Distributors (MVPDs) connected the majority of television households to cable television and ultimately broadband Internet, the shelf space for audio video content was largely limited to broadcast signals (TV and radio stations) and physical media (cassettes, phonographs, discs). As a greater percentage of TV households became connected to cable television, the shelf-space for TV content took off.

First with audio recordings and then with video, the digital revolution opened up vast new shelf space for content, lowered the cost and barriers to entry for new competitors in this space, changed the entire nature of consumer advertising, and ultimately disrupted the business models for the traditional media companies that long-dominated this space.

No longer limited by the number of networks carried by their MVPDs, American consumers now had access to millions of hours of video content streamed through YouTube. Some of this content was crude user-generated video; some was polished, professional, and often illegally posted premium content; increasingly, much of the content was carefully crafted semi-professional content that sometimes featured professional-level production values and was simply posted by creators who aspired to express themselves through the medium of moving images.

Capitalizing on streaming technology, "Over-the-Top" platforms such as Hulu, Netflix, and Amazon Prime started offering consumers other ways to access content, and increasingly offered exclusive original programming. Awash with a wide variety of content and original programming, American consumers grew to expect to access content that specifically appealed to them, when they wanted it, where they wanted it. With the diversity of audience tastes and the vast shelf-space, audiences fragmented to degrees not seen since the advent of recorded media. Gone were the days where large segments of the population regularly tuned into watch the same content.

As a result, both new and established media platforms must aggressively compete to build an audience. At the same time, the Internet is making it easier for audiences to access pirated content.

The net effects of these forces have resulted in three meaningful industry changes. The first is increased competition among producers for creating quality content. The second is the growth of niche content. The third, which in many ways results from the first two, is an increase in the aggregate global demand for American content.

7.3 Demand for Labor

So how do the changing consumption patterns impact the demand for labor in the media sector? The production of niche-audience content has had two meaningful impacts. More content production, even if intended for small audiences, drives up the number of jobs (roles). However, this content often has a shorter lifecycle, which means there is less of an opportunity for back-end (residual) payout. At the same time, competition for audiences increases demand for high-budget content, which bids up the wages of the top performers that are in the highest demand.

7.4 Supply of Labor

A major impact of the Internet on this sector is that it lowers barriers to entry. Whereas in the past television networks and other major media properties controlled access to audiences, the Internet has made it much easier for both producers and performers to reach audiences directly. For talent, this means it is much easier to "break into the business;" anyone can become a media personality now, and it's not hard to monetize this work, even if to a minimal extent. More people see a career on camera or behind a microphone as a realistic career to aspire to. Many of these people are also more likely to continue to pursue this career, even if it takes them a while. Essentially, being a "media professional" is less of a black-and-white distinction than it was in the past. These factors have served to drive up the overall supply of labor in the sector.

7.5 Aggregate Employment

As may be expected by the changes in supply and demand for labor, the Internet appears to be a net job creator for performers. However, while positive, the impact is still relatively minimal. By 2014, about 3% of all SAG-AFTRA covered audio-visual entertainment jobs were for work in digital media (SAG-AFTRA 2016) (Fig. 7.1).

Furthermore, the newly created jobs in digital media are far from uniform. The majority of these new jobs are for roles in low-budget productions with small relative audiences. However, more and more of this work is done for original programming distributed on streaming services such as Netflix—programming that more closely resembles the higher-budget content produced for traditional pay TV services. While original digital video content has been produced for quite some time, it is only very recently that this type of content has been produced at budget levels comparable to traditional television.

Interestingly, these new higher-budget digital productions are apparently not cannibalizing traditional television work, and appear to be in some ways fueling the growth of traditional television production. As traditional television platforms face increased competition from streaming video platforms, their reaction has been to produce and order more original programming in order to retain viewers and subscribers. So some of the increase in jobs in traditional television may be in part due indirectly to competition from the Internet programming.

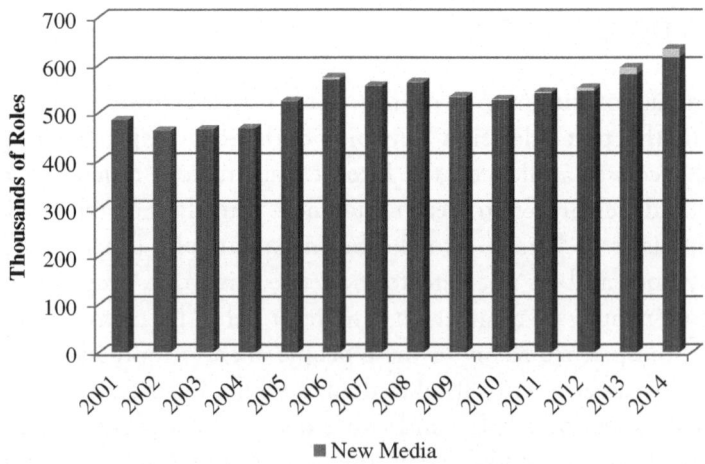

Fig. 7.1 SAG-AFTRA-covered audio-visual entertainment jobs

7.6   Performer Income

The impact of the Internet on performer income is a somewhat more complex issue. On one hand changes in supply and demand have generated some upward pressure. As new exhibition platforms have been opened (subscription streaming services, electronic-sell-through, etc.), more and more library content can be distributed. This increase has worked to drive up aggregate residuals earnings, as performers are being compensated with new residuals streams. Also, competition among producers has bid up the compensation for "star" performers. As the media landscape becomes more crowded, many producers are willing to invest top dollar to capitalize on the popularity of celebrities as a way to cut through the media clutter.

However, at the same time, there are forces at play creating downward pressure on performer income. With more performers in the workforce, increased competition for jobs exerts downward pressure on earnings for non-"star" performers. Since content now tends to have a shorter shelf life, the average residuals payment per job worked tends to be lower. The net effect of these changes is largely a further migration toward a "Winner-Take-All" labor market. Those workers at the high-end of the income spectrum are able to earn even more, while most other workers are seeing little-to-negative increases in their net earnings.

7.7   Benefits and Drawbacks for Labor

This further migration toward a "Winner-Take-All" labor market has various meaningful implications for professional performers. One substantial upside is an increase in the net number of jobs available to performers; another is the increase in aggregate compensation. For a select few "stars," it has yielded even greater compensation. However, these changes bring significant downsides for performers as well. Professional performers on an individual basis are now seeing greater income variability from year to year, resulting from increased competition and lower average residuals. This ultimately means that its harder for many performers to maintain and sustain a career in the business.

7.8   Protecting Performers

As the labor union representing these performers, SAG-AFTRA is responsible for protecting these often-vulnerable workers. SAG-AFTRA uses the leverage of collective bargaining to ensure that performers in these sectors receive fair initial compensation. Further, we negotiate to ensure that performers are paid fair residuals when their work is exhibited on emerging distribution platforms. Ultimately, these protections enable our members to continue to earn a living in a highly volatile and competitive industry.

7.9   Conclusions

The Internet has undoubtedly changed the labor market for talent in the media business. The Internet's impact on the sector is still evolving, but some insights can already be drawn. The Internet has added new jobs for performers; however, the vast majority of these new jobs tend to be at the lower end of the pay spectrum. The Internet has also created new ways to distribute and monetize library video content. While these new revenue streams are becoming more significant in the aggregate, they have, however, yet to become a significant source of income for individual performers. Furthermore, the Internet has disrupted this labor market by driving up competition and year-to-year income variability. At this point it is unclear whether the incremental gains in lower-paying jobs and new, but smaller, residuals payments are meaningful enough to outweigh the increased competition, wage compression, and income variability that professional performers now face in the Internet age.

References

SAG-AFTRA. 2016. "SAG-AFTRA Analysis of Session Earnings Reported to the SAG-P Pension & Health and AFTRA Health & Retirement Plans." Earnings Analysis.

Vogel, Harold L. 2014. *Entertainment Industry Economics: A Guide for Financial Analysis*. Cambridge: Cambridge University Press.

Part II

Internet Economic Fundamentals and Their Impact on Economy and Distribution

8

Inequality and the Digital Economy

Eli Noam

8.1 Introduction

For many years, policymakers in developed countries have believed and hoped that the Internet, and more generally the digital economy, would replace and enhance industrial jobs. This was important to developed countries as their traditional manufacturing activities were either being automated or were migrating to developing or emerging countries. It was also important as a way to find a productive space for younger generations who moved from the blue-collar jobs of their parents to knowledge-based occupations where they could utilize society's investment in their higher level of education. Lastly, such jobs, with their replacement of low-paying factory drudgery with well-paying digital tasks, were also considered to reduce class division and inequality.

Thus, in countries undergoing deindustrialization—and which isn't, among developed economies—an Internet-based economic growth has been widely recommended as a way to create economic activity and reduce the inequality of postindustrial society. In particular, the opportunities that the Internet affords to the "creative workforce" are believed to be an engine for employment.

Because this is important we have to confront the question openly, with no nostalgia for the past, with eagerness for the future, but without wishful

E. Noam (✉)
Columbia Institute for Tele-Information,
Columbia University Business School, New York, USA

© The Author(s) 2018

117

L. Pupillo et al. (eds.), *Digitized Labor*, https://doi.org/10.1007/978-3-319-78420-5_8

thinking about the present. We now had several decades of digital evolution, and it becomes possible to measure rather than postulate. So let us look at the Internet as a job creator.[1]

8.2   Gains

The conventional story is one of great success. Industries that are negatively affected tend to be viewed as inefficient oligopolies such as the music industry or daily newspapers. The Internet is supposed to have caused up to 21% of GDP growth in 5 years in mature countries.[2] In the US, the Internet economy has reportedly created 1.2M jobs[3] directly, according to a Harvard Business School study.[4] In France, too, the Internet has supposedly created 1.2 million jobs directly.

But what kind of jobs? In the US, most of them were in e-commerce, not in anything really creative, but mostly in order fulfillment, i.e., packaging and shipping, as well as the delivery of the physical goods such as through trucking, accounting for more than 500,000 of the 1.2 million jobs. Internet service providers generated 181,000. Creative jobs were, in particular, in content-related employment, estimated at 60,000, and in software as a service, 31,500.[5]

These modest numbers are in contrast to the sometimes breathless hype. For the Internet of Things, a trade magazine gushed "While today there are just 300,000 developers contributing to the IoT, a new report from VisionMobile projects[6] a whopping 4.5 million developers by 2020, reflecting a 57% compound annual growth rate and a massive market opportunity."[7]

There were also new types of jobs spawned by various applications such as, new taxi drivers[8] due to the car service app Uber, or by people creating their own new income streams, for example, by renting out driveways on Parking Panda.[9]

There are also many indirect job creations. A study found that each Internet job supports approximately 1.54 additional jobs elsewhere in the economy.[10] Another study, conducted by the McKinsey Global Institute on data from 13 countries, found that for every job destroyed, the Internet created 2.6 new jobs, for a net addition of 1.6.[11]

8.3   Losses

Let us now also look at the downsides. Even a technology booster, Bill Gates, warned

Software substitution, whether it's for drivers or waiters or nurses … over time will reduce demand for jobs, particularly at the lower end of skill set. … 20 years from now, labor demand for lots of skill sets will be substantially lower. I don't think people have that in their mental model.[12]

Gates was half right. Some job losses are upon us, as one would expect in any transition. But are they at the lower end of employment or at the middle?[13]

8.3.1 The Internet-Induced Job Losses in the Industrial Sector

In the US, industrial blue-collar jobs have disappeared at the rate of 350,000 industrial jobs each year in the US for 2 decades now. Plus the multiplier effect of jobs, about 1.6 per industrial worker and 2.5 per skilled industrial worker. This adds up to a job loss of about half a million each year.[14] Of course, many would have disappeared anyway, but more slowly. Transition time is important. People would have had more time to adjust, retrain, and relocate. The Internet has accelerated the outmigration of jobs. Erik Brynjolfsson and Andrew McAfee of MIT argue in their book, *Race Against the Machine*,[15] that progress in information and communication technology (ICT) may be occurring too fast for labor markets to keep up.

Take the photography company Kodak. It employed more than 140,000 people. It even invented the first digital camera. But Kodak went bankrupt when that same digital photography moved to Internetworked mobile phones. A major player in that new field is Instagram. Instagram was bought by Facebook for a billion dollars in 2012, and reportedly employed only 13 people.[16] It, and the designers of digital cameras will not provide employment for the tens of thousands of Kodak manufacturing workers who lost their jobs. Such digital camera makers, in the US, include General Imaging, Cobra, and SeaLife. Best known is the action-oriented GoPro. Its products, however, are built in China and Brazil.

8.3.2 The Impact of the Internet on Service Jobs

Following the blue-collar jobs, the pink-collar jobs in retailing and clerical staffs began to shrink as retailing moved online. Similarly, service support jobs such as telemarketing or editorial work have been moving offshore. Middle management levels have been cut as ICT made supervision and information exchange easier, thus reducing the need for intermediate levels of management.

Online shopping has been growing steadily, with a US share of above 12% ($473 billion) of total retail (4.03 trillion) and rising.[17] In the UK, a research project predicted, according to its director, professor Joshua Bamfield: "We expect 4000 to 5000 stores to close due to competition from online retail, with an acceleration in chains closing stores to focus more on online operations."[18] In America, the drop in retail jobs since 2007—after a four-year boom—has been pronounced, with a reduction of 900,000 jobs in 5 years, a nearly 6% decline.[19]

Retailing is not the only service industry to be affected. A short list of some of the major industries destroyed by the Internet[20] includes newspapers, travel agencies, stock brokers, and soon also universities. So we have more than a deindustrialization, but also a "de-servicization."

8.3.3   The Unequal Impact on Different Income Classes

The problem is not just the loss of traditional employment at a pace that was hard to counteract by digital employment, but that the losses are distributed unequally. In the United States, half the 7.5 million jobs lost during the Great Recession were in industries that pay middle-class wages. But only 2% of the 3.5 million jobs gained since the recession ended in June 2009 were in midpay industries. Nearly 70% are in low-pay industries, and 29% in industries that pay well.[21]

In the 17 European countries that use the Euro as their currency, the numbers are even worse. Almost 4.3 million low-pay jobs have been gained since mid-2009, but the loss of midpay jobs has not stopped. In Japan, a 2009 report from Hitotsubashi University in Tokyo documented a "substantial" drop in midpay, midskill jobs in the five years through 2005, and linked it to technology.[22]

Many middle-level jobs are easier to automatize by smart software programs, or to outsource and offshore, than low-level jobs. One can automatize travel agents and bank tellers, but it is harder to do it for road construction or cleaning crews.

A study by David Autor of MIT and David Dorn of the Centre for Monetary and Financial Studies in Madrid[23] graded occupations in terms of their vulnerability to automation. They identified the jobs of secretaries, bank tellers, and payroll clerks as among those most dominated by routine tasks. Industries that adopted IT at faster rates also saw the fastest growth in demand for the most educated workers, and the sharpest declines in demand for people with intermediate levels of education. Thus, whereas in the 1970s and 1980s employment in middle-skilled, middle-income occupations grew faster than that in lower-skilled jobs, by the 1990s employment in middle-class jobs began to decline as a share of the total while the share of both low- and high-skilled jobs rose.

The data shows these trends, with middle-income occupations losing out, while upper and lower income occupations have been gaining (Fig. 8.1).

Of course, warnings about challenges to the middle class have been around for a long time, as Robert Gordon has pointed out,[24] and been a staple for political candidates on the stump. ICT was seen as a way to turn this around. Yet a study by Guy Michaels, Ashwini Natraj and John Van Reenen of the London School of Economics for 11 countries finds that industries that adopted ICT at faster rates (as measured by their spending on ICT and R&D) also experienced the fastest growth in demand for the most educated workers, and the sharpest declines for those with intermediate levels of education.[25]

This "hollowing out" of the middle-class workforce will continue.[26] The US Bureau of Labor Statistics predicts that employment in low-skilled service occupations will increase by 4.1 million, or 14%, between 2008 and 2018. The only major job category with greater projected growth is that of

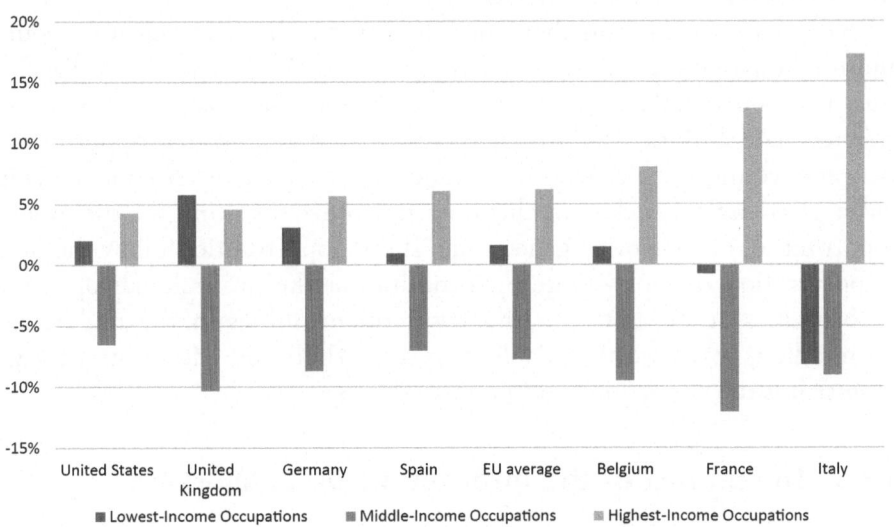

Fig. 8.1 Change in Employment Shares, 1993–2006 (*Source* Data from Goos, Maarten, Alan Manning, and Anna Salomons. 2009. "Job Polarization in Europe." *American Economic Review* 99 (2): 58–63)

professional occupations, which may add 5.2 million jobs, or 17%.[27] It is much lower for middle-class jobs. So we create a bottleneck. Menial Jobs at the bottom, professional jobs at the top, and a weakening in the middle.

This has a lot of implications. It means that the job mobility from lower to middle class, which had been the historic way to individual progress, is becoming more difficult. The lower occupations are blocked. Social mobility is thus declining. For much of the twentieth century, people's job prospects rose with extra education. While this is still true, the effect is lessened at the lower end. And this happens at the time when the cost of education keeps climbing steeply.

8.3.4 The Impact of the Internet on Younger Workers

It is generally believed that the Internet leaves behind older folks unprepared for the digital age, but that it is a great improvement in the opportunities of young people. If so, how come their standard of living today is lower than those of the preceding generation, and how come there is such a huge youth unemployment in many advanced countries?[28] Youth unemployment in 2015: USA 11.7%,[29] United Kingdom 14.4%,[30] France 24.1%.[31] If the Internet has done all these great things for the digitally native generation, and if it has made distance obsolete, how come they live more than ever with mom and dad?

There is a great illusion that since the Internet has been creating young multi-billionaires like Mark Zuckerberg, Sergei Brin, and Larry Page, it must be good therefore for an entire generation. But that is a sloppy and incorrect conclusion. The Internet creates, indeed, greater opportunities for a few young people, who have education, a spirit of entrepreneurship, and a great deal of luck. But this does not prove anything for the average opportunities of the young generation. Those opportunities follow the overall polarization of jobs—more opportunities at the professional top. More opportunities at the bottom. And fewer opportunities in the middle. Yet the middle is where young people with some skills and education must go, because it is the classic jump-off point to the top.

8.3.5 The Impact of the Internet on Older Workers

Paradoxically, a similar problem happens at the other end of the age spectrum. The rapid change in knowledge and technologies means that the learning curve is short, and that there is less value to the experience.

In the past, an experienced elder had advantages. Now, the old become expensive, out of date, and expendable. They get bumped out of the middle level jobs where there is less room and the competition for the jobs is tougher than before. Their skills become obsolete for the top jobs. And the medial jobs at the bottom are often physically too demanding. So there is less room for older workers. And this is just at the time when life expectancies rise. When retirement systems become unaffordable to societies. And when companies find ways to avoid paying taxes to contribute to the pot. Thus, the same technological progress that enables society to keep old folks' bodies alive longer is also shortening the value of their minds.[32]

8.3.6 Is the Creative Sector the Remedy for These Job Losses?

Is the creative sector going to be the substitute for all of those industrial and service sector jobs that are being lost? This claim, often heard, is absurd if one looks at the numbers. It just shows that many creatives who make the argument, or politicians and intellectuals, substitute wishful thinking for statistics. In America, the number of industrial jobs lost over the past decade has been 5 million, including the multiplier effects.[33] The number of retail jobs lost has been over a million.[34] The number of people with jobs in journalism, books, TV, film, theater, music, is less than one million.[35] So if creative jobs alone should do the compensation one would have to expand that sector by a factor of 7. Who would watch, read, or listen to all this new creation? People are not going to watch 7 times more TV, they already watch 7 hours a day. Plus, a lot more people produce content as volunteers, not as a job. On top of that, the globalization of media means that every other country's content is also available, and is also expanding, by the same logic, by a factor of 7. And, who is going to pay for all this, so that these creators actually get a paycheck?

While the claims of creative jobs that will offset industrial and service losses are being touted, let us take note that journalist jobs are melting like butter in the sun.[36] That most musicians do not get paid anymore by anyone. That fewer people read books, though more books get written. That TV networks are a shade of their former selves, and the cable networks are very leanly staffed operations.

Even if one expands the definition of creative to software and design, and even if one includes a generous multiplier, the numbers do not support the notion that the creative sector will be an offset to the industrial and service job losses.

To conclude: the Internet is a force for inequality. It creates inequality among occupational classes, among regions, and among generations.

The preceding discussion can be equally adopted by neoliberal economists on the right and critical media scholars on the left. Where they part company is in the analysis of the causes, and therefore also of the remedies. Many of the latter believe that the problem is caused by profit-focused Internet moguls. Stop them and the world will be a better place. But this view is quite wrong even in terms of critical analysis. Some people end up at the top, by luck, pluck, and connections. The real question is not who ends up at top but whether the new technology defines economics, which defines a market structure and an employment configuration whose equilibrium is socially objectionable.

Thus, the emerging unequal employment system may well be not the result of failure but of success. It is the result of fundamental economics that restructure economies fundamentally. And because they are fundamental they are very hard to deal with through government policy.

8.4   The Fundamental Economics of the Digital Economy

8.4.1   Fundamental Characteristic #1 of the Digital Economy: Digital Activities Are Typically Characterized by High Fixed Costs, Low Marginal Costs, and Network Effects

Do the new Internet media make a difference on media industry concentration, in the way its enthusiasts believe? Internet media, after an early stage of a dynamically competitive market structure, often becomes highly concentrated. Various market segments have their dominant players—Amazon, EBay, Microsoft, Google, Facebook, Twitter, YouTube, Apple's App Store, and others. The Internet sector was believed to be wide open and competitive and would open things up for other industries, but it exhibits strong concentration trends. The underlying economics on the supply side are, high fixed cost and low marginal cost; and on the demand side, strong network effects. This leads to major economies of scale. And therefore, it results in highly concentrated industries, with a few firms the winners.

New information industries are more capital intensive than old ones. Their ratio of capital costs to operating costs is higher than in the past. In business terms, capex was growing while opex was declining. In consequence, their scale economies are greater and their market concentration is higher.

There are several business implications. The economies of scale lead to large-sized companies and consequently to a high market concentration. In the extreme, one encounters a winner-takes-all near-monopoly. There are therefore incentives to reach large size through mergers and/or to be a first-mover in a product in order to gain scale.

It means that the market structure among companies is highly unequal. Some firms win big, most lose out or are weak. This trend is likely to continue, especially if the pace of disruptive innovation in the sector slows down a bit.

8.4.2 Fundamental Characteristic #2 of the Digital Economy: A High-Risk Distribution of Success

A major characteristic of media is its high risk in the presence of competition. One often observes a "80–20" outcome in which 80% of all media products do not become profitable, 90% of all profits are generated by 10% of the products, and 50% of profits are generated by 1–2% of products.

Every industry and company is structured like a tournament. And the question is how such a tournament is set up. Is it "winner-takes-all"? The economic literature tells us that the higher the risk in the tournament, the greater one must make the disparity between the winners and the losers. One must compensate the players by a higher jackpot.[37]

In accordance with this analysis, creatives' incomes are much more unequally distributed than regular incomes, due to the risk characteristics of their companies and industries. And that can be observed. The tournament profile of compensation for aspiring creatives is extraordinarily steep. Pay differentials in media are especially high due to an oversupply of talent, as well as due to an incentives structure where the few "winners" receive the majority of the reward.

Creatives usually overestimate the odds for personal success.[38] They also accept low compensation and high risk because of the high level of personal satisfaction inherent in artistic careers. In creative activities, such as film and TV, or in sports, small differences in talent may typically result in extreme differences in reward.[39] These small talent differences are rewarded exponentially rather than linearly, which leads to highly skewed distribution of rewards. This model applies to many industries, but it is most pronounced in the creative industries because spots at the top are scarcer, and the bottom is much wider.[40] Thus, an economy with a stronger participation by creative is a more unequal economy. And a digital economy with a strong reward system for the winners is more unequal.

8.4.3 Fundamental Characteristic #3 of the Digital Economy: The Presence of Non-Maximizers of Profit

Many individuals in the digital field derive utility from the process of creating a product, not from profiting from its sale. Producing the good is not a chore but a benefit. When this occurs it is hard to distinguish production from consumption. In media production, creatives are incentivized to maximize recognition, not profit, or a combination of the two. Online media provide a greater way to create content and find an audience by lowering the cost of production and distribution, and hence have increased this non-profit participation. As a result, it becomes more difficult for participants to survive economically. The notion that "one can't compete against free" affects companies as well as individuals. This is another factor in skewing the income distribution further.

8.4.4 Fundamental Characteristic #4 of the Digital Economy: Excess Supply

Media production increases exponentially at a substantial rate, while media consumption increases linearly and slowly. Content rises by about 12%, and attention rises by less than 4%. Given the gap between production (supply) and consumption (demand), excess supply is inevitable. Compared to 1998, fewer than half as many of the new products make it to the bestsellers lists, reach the top of audience rankings, or win a platinum disk.[41]

In almost any scientific field, more research articles were written just this year alone than in the entire history of human beings before 1900. In the field of chemistry, within a span of thirty-two years (1907–1938), one million chemistry articles were authored and abstracted. In contrast, it took less than one year for a million such articles to be produced in 2010. Every thirty seconds, a new book is published. Every day, ten new feature films and 1500 television shows are produced.

This has consequences for both content style and marketing.[42] Attention is the scarce resource.[43] As Nobelist Herbert Simon observed, "a wealth of information creates a poverty of attention."[44] New media consumption must be mostly supported by substitution from existing media in terms of time or full attention. Inevitably, this leads to rising competition for "mindshare" and "attention." Such competition, in turn, leads to pressure on prices, discussed in the next section, and hence to a lowering of compensation.

8.4.5 Fundamental Characteristic #5 of the Digital Economy: Price Deflation

When competition occurs, prices drop toward long-term marginal cost. In the short term, marginal cost is near-zero and does not cover fixed cost. The result of price competition with low marginal cost has been price deflation in information products and services. This is a good deal for the consumer who enjoys substantial "consumer surplus." They must shell out much less than they would be willing to pay if they had no choice. But it creates a difficult problem for the supplier.[45] Price deflation to marginal cost poses a threat to long-term viability since low prices make it difficult to cover costs and achieve profitability. And that is indeed what has been happening. Information has become cheaper for many a decade. It is often becoming difficult to charge anything for it. Music and online content is increasingly free. Free newspapers are being handed out. Such a price deflation is one of the fundamental economic trends of our time.[46] The entire competitive part of the information sector—from music to newspapers to telecoms to Internet to semiconductors and anything in-between— has become subject to a gigantic price deflation in slow motion. This leads to economic pressure, to price wars which squeeze out weaker companies and subsequent consolidation of the more viable survivors, and to lower compensation for all employees.

8.4.6 Fundamental Characteristic #6 of the Digital Economy: The "Reverse" Cost Disease

For a long time, the income in creative industries has risen, even though productivity has not. This impact is known as the "cost disease," a term coined by William J. Baumol and William G. Bowen.[47]

The "cost disease" phenomenon seems counterintuitive. In the long run, workers' real incomes rise due to their rising productivity. This raises incomes across the economy. One must therefore pay low productivity occupations, like creatives in media, more than before, because they now have better-paying alternative opportunities. These increases in the cost of production may offset the cost savings from any technical progress in those creative activities. Thus, workers in occupations experiencing no growth in labor productivity at all nevertheless receive higher wages as a result of increases in productivity in other sectors of the economy. The labor-intensive performing arts thus become relatively costlier to produce, thus showing low productivity.[48] And yet, the people employed in these activities actually get paid more than in the past.

However, this driver of income growth characterized the past. The same logic now depresses creatives' incomes. As industrial wages decline relatively, they also affect the creatives' compensation.

8.4.7 Fundamental Characteristic #7 of the Digital Economy: Instability

As a result of the various factors, the digital economy is more volatile than the industrial economy. It is more subject to economic cycles and greater instability. The dynamics are as follows: an innovative idea raises hope. A boom gets on its way, becoming a bubble. But in a competitive environment, competition drives prices down to marginal cost. Marginal cost is close to zero. Such a price is not sustainable for long. Companies go out of business en masse. Investors flee. The economy descends into a downward spiral. But soon, the survivors stabilize the industry. Prices rise, and with it profitability. At that point, new entrants emerge. The industry becomes more competitive. A new cycle begins.

Thus, the information economy is an unstable economy. And because of the acceleration of technological progress ("Moore's law"), the cycles almost inevitably accelerate in frequency and maybe in amplitude.

8.4.8 Fundamental Characteristic #8 of the Digital Economy: The Transformation of Firms into "Network Companies" Leads to a "Freelance" Economy

The economic system based on the electronic networks changes work relations. Firms become organized as networks. They hire by project. They outsource to contactors. They do everything they can to reduce the fixed costs and to shift it to others. Examples are computer chip making and film production. Most chips today are designed by companies but not manufactured by them. Sometimes even the design gets outsourced to design bureaus. The same holds true for the Hollywood studios. Most of the films they distribute are made by independent entities, which in turn contract with others for their temporary services. Increasingly, collaborators are assembled for projects on a project basis. Companies contract workers, consultants, and outsourced vendors.[49] In the same way that "just in time" production has shifted manufacturing, capital assets, inventory, and risk to the suppliers of components, so it is now giving rise to "just in time" workers—employees whom a business can hire on a moment's notice to fill a moment's need.[50] These "just in time" workers have few of the benefits that traditional employees have gained over time. No health and safety protections, retirement plans, or overtime pay. This organizational model has the potential to become the model for the mainstream firm of the future, given its project-oriented, fluid management structure, flexible skills deployment, and reduction of fixed costs.

8.5 Consequences for Digital Management

Is it realistic for digital managers to think that they can avoid these issues? That they can take credit for just about anything positive that is happening, from Tahrir Square to the Obama election to microfinance in developing countries, but that somehow the negative developments are someone else's fault?

There are two tracks for digital companies' actions, the first is that of their policy positions, and the second is that of management actions. As profit maximizing managers, they will inevitably create value and wealth, but also be part of creative destruction. Outside of noble but superficial philanthropic and socially responsible actions, they cannot avoid the fundamental forces described above. This means that they must understand the environment in order to function in it.

8.5.1 Expect a Return of Unionization

To create employment benefits for the new type of employees, labor unions in the freelance tech sector are likely to emerge, following the model of unionization of creatives in theater, film, and music. The constraint is the difficulty of effectively striking when the work can be easily outsourced to offshore locations. This suggests that the most likely strategy of labor will be that of political pressure and legislation.

8.5.2 Expect a New Wave of Political Disputes and Activism in the Digital Economy

Income and employment issues are part of a much larger discussion over the control of information resources. This includes advocacy for unimpeded access of content to the Internet ("net neutrality")[51]; the "open source" movement that battles Microsoft[52]; the "copyleft" community that challenges copyright systems that favor media companies[53]; the privacy protection advocacy against the use of personal information by marketers and governments[54]; the peer-to-peer file sharing, which has moved beyond financially convenient piracy to an ideology on cultural creativity[55]; the "open innovation" concept of user-based technology communities that has challenged the traditional proprietary R&D system[56]; the "unlicensed spectrum" initiatives that seek to undermine the exclusivity of access to airwaves of broadcasters and wireless providers[57]; the push against a "digital divide" that is based on the income, skills, and geography[58]; the move to municipal and free WiFi connectivity challenging phone companies[59]; and more. All of these developments have their particular reasons but also a common thread. They are manifestations of a wider conflict over the extent and nature of control in the information society.

Most observers are familiar with the various flash points but have not always connected the dots and recognized the emerging social movement on the model of environmentalism. For years, information companies and governments have touted their activities as the key to the planet's economic and cultural future and the solution to most of its problems. No wonder that control over this sector is being contested by more than business competitors. As the information sector permeates society, society, in turn, permeates the information sector with its internal and international conflicts.

8.5.3 Expect a New Wave Government Policies

Given these fundamental economic and technological drivers, it is almost inevitable that the economic equilibrium of the internet economy, left to itself, will not be at a level of diversity and employment that many people consider necessary. Recent decades have led to a reduction of regulatory restrictions and interventions because of the expectation that technology and market forces would overcome inequalities in democratic societies. If this hope is not realized, the pendulum will inevitably swing back to various interventionist approaches of regulation, breakups, and subsidization, promoted by the various activist initiatives described above.

But governmental actions are becoming more difficult. Government rules worked moderately well on the operational level when industries were simple and tools of control existed. But the government's powers today are much more limited. If Google has significant market power in Argentina, how should or could the search engine market there be restructured? If a Korean firm is dominant in interactive games, what then is the Swedish (or the EU's) government's remedy? If Skype's voice quality declines, who would deal with that, if at all, and how? And these are merely conceptual questions, to which are added those of politics, litigation, international trade, intellectual property rights, and international enforcement. It is always difficult for laws or regulations to modify fundamental transitions of industries. It is particularly difficult to do so where, as in the case of media infrastructure of the Internet, any policy in a free society needs to be done with a light touch.

8.5.4 The Need for Direct Business Action

Digital firms need to contribute directly to overcoming some of the dislocation effects which they have created, or else they will find themselves regulated in unfavorable ways. This goes beyond a little PR-driven philanthropy. They should channel their talent, creativity, and problem-solving skills to help those on the losing side of the equation. On the educational and skill side, they should contribute to STEM education, and send some of their best and brightest team members to teach it. This should include the training of older people. It is disgraceful how the Internet sector first marginalizes older folks, then makes fun of them, leaves them destitute, and does not lend a helping hand.

8.5.5 Stop Claiming to Be the Solution

The least productive approach for digital managers is to try to take advantage of a problem they helped create, by proclaiming that they should receive more help, more governmental support, more access, less regulation, etc. and everything will be fine. That will only damage their credibility and backfire.

8.5.6 Stop Arguing That Seeing a Problem Is Anti-technology Luddism

Every time there is a technology shift, there are doubts and fears. Throughout history, technology has been a job creator.[60] But that did not help those that were dislocated. In the Industrial Revolution, which proceeded at a much slower pace, millions of Europeans ended up destitute and had to migrate to sprawling city slums or to distant shores. Social and political revolutions and upheavals abounded. Now, the pace of dislocation is even faster. And the problem might be deeper. As the MIT study by Brynjolfsson and McAfee argues:

The pattern is clear: as businesses generated more value from their workers, the country as a whole became richer, which fuelled more economic activity and created even more jobs. Then, beginning in 2000, the lines diverge; productivity continues to rise robustly, but employment suddenly wilts. By 2011, a significant gap appears between the two lines, showing economic growth with no parallel increase in job creation.[61]

8.5.7 Recommend, Support, and Finance Governmental Actions Where Appropriate

What might such governmental strategies be? This is a big topic. Foremost, it should become easier for people to retrain and obtain new skills. One proposal: let people of over 40 years of age take time off—say a year—to retrain. This would be funded—tuition and living expenses—by the Social Security system (or similar pension arrangements elsewhere). In return, the retirement benefits for the person would commence later. For each month of retraining "sabbatical," that person's retirement date would go up by one month. This would approximately fund such a system. There should also be a required job training during an unemployment period where public benefits are received. There should be free online or part-time training courses with certifications. And there should be mandatory and higher requirements for STEM courses in both secondary schools and college, and more vocational training. College degrees should have required dual majors: one in a "practical" area, and the second one in any other area.

A second major governmental strategy is to strengthen those industries that are more stable. These tend to be local in nature rather than global. They are, in particular, industries that deal with the human body, with the home, with maintenance, and with hospitality and leisure. Some jobs are still beyond the reach of automation and outsourcing: construction jobs, jobs with unpredictable patterns, jobs requiring dexterity and judgment, jobs for small batches or with numerous variations,[62] or jobs that require an understanding of human nature. Such jobs, as well as those that are local in nature, deserve development priority.

To conclude: The impact of Internet-induced economic displacements in developed economies will not go away. It will get worse in the short term. This creates a challenge to managers and policymakers in the digital sector. Otherwise, a backlash will create forces that will restrict innovation. It is therefore important for academics, public–policy analysts, NGOs, companies, and governments to think creatively about new approaches to these issues, and to balance the public interest, technological innovation, and financial investment in the emerging environment.

Notes

1. An earlier and preliminary version of this article is in the Pamplona conference proceedings of the International Media Management Academic Association, Wildman, Steve, and Monica Herrero, eds. (2015), *The Business of Media: Changes and Challenges* (Lisbon: Media XXI).
2. Du Rausas, Matthieu Pélissié et al. (2011), "Internet Matters: The Net's Sweeping Impact on Growth, Jobs, and Prosperity," *McKinsey Global Institute*, May, https://www.mckinsey.com/industries/high-tech/our-insights/internet-matters.
3. Thibodeau, Patrick (2009), "Study: Internet Economy Has Created 1.2M Jobs," *Computerworld*, June 10, http://www.computerworld.com/article/2525229/internet/study–internet-economy-has-create-1-2m-jobs.html.
4. Quelch, John (2009), "Quantifying the Economic Impact of the Internet," *HBS Working Knowledge*, August 17, http://hbswk.hbs.edu/item/6268.html.
5. Thibodeau, Patrick (2009), "Study: Internet Economy Has Created 1.2M Jobs," *Computerworld*, June 10, http://www.computerworld.com/article/2525229/internet/study–internet-economy-has-create-1-2m-jobs.html.
6. VisionMobile (2014), "IOT: Breaking Free From Internet and Things: How Communities and Data Will Shape the Future of IOT in Ways We Can't Imagine," *VisionMobile*, June, http://www.visionmobile.com/product/iot-breaking-free-internet-things/.
7. Asay, Matt (2014), "The Internet of Things Will Need Millions of Developers by 2020," *Readwrite*, June 27, http://readwrite.com/2014/06/27/internet-of-things-developers-jobs-opportunity.
8. Bensinger, Greg (2013), "Apps Are Creating New Jobs," *Wall Street Journal*, March 5, http://online.wsj.com/news/articles/SB10001424127887323864304578320861732248742.
9. Ibid.
10. Quelch, John (2009), "Quantifying the Economic Impact of the Internet," *HBS Working Knowledge*, August 17, http://hbswk.hbs.edu/item/6268.html.
11. Du Rausas, Matthieu Pélissié et al. (2011), "Internet matters: The Net's Sweeping Impact on Growth, Jobs, and Prosperity," *McKinsey Global Institute*, May, https://www.mckinsey.com/industries/high-tech/our-insights/internet-matters.
12. Bort, Julie (2014), "Bill Gates: People Don't Realize How Many Jobs Will Soon Be Replaced by Software Bots," *Business Insider*, March 13, http://www.businessinsider.com/bill-gates-bots-are-taking-away-jobs-2014-3.
13. Ibid.
14. Atkinson, Robert D. et al. (2012), "Worse Than the Great Depression: What Experts Are Missing About American Manufacturing Decline," *The Information Technology & Innovation Foundation*, March, Last Accessed September 14, 2015, http://www2.itif.org/2012-american-manufacturing-decline.pdf.

15. Brynjolfsson, Erik, and Andrew McAfee (2011), *Race Against the Machine* (Lexington, MA: Digital Frontier Press).
16. Timberg, Scott (2013), "Jaron Lanier: The Internet Destroyed the Middle Class," *Salon*, May 12, Last Accessed July 23, 2015, http://www.salon.com/2013/05/12/jaron_lanier_the_internet_destroyed_the_middle_class.
17. Mintel. "Market Sizes," Last Accessed June 23, 2015, http://marketsizes.mintel.com.
18. Reilly, Jill (2013), "Booming Internet Sales 'Will Close 5,000 High Street Stores and Cost 50,000 Jobs'," *Daily Mail*. January 1, http://www.dailymail.co.uk/news/article-2255677/Booming-Internet-sales-close-5-000-High-Street-stores-cost-50-000-jobs.html.
19. Wright, Joshua (2012), "The Demise of Retail Jobs? Not So Fast," *emsi*, April 16, http://www.economicmodeling.com/2012/04/16/the-demise-of-retail-jobs-not-so-fast.
20. Briefing Investor (2012), "Industries Destroyed by the Internet—A Reflection," July 26, http://www.briefing.com/investor/our-view/ahead-of-the-curve/industries-destroyed-by-the-internet–a-reflection.htm.
21. Condon, Bernard, and Paul Wiseman (2013), "Millions of Middle-Class Jobs Killed by Machines in Great Recession's Wake," *Huffington Post*, January 23, http://www.huffingtonpost.com/2013/01/23/middle-class-jobs-machines_n_2532639.html.
22. Ibid.
23. Autor, David (2010), "Polarization of Job Opportunities in the U.S. Labor Market," *Center for American Progress and The Hamilton Project*, April.
24. Gordon, Robert J. (2016), *The Rise and Fall of American Growth: The U.S. Standard of Living Since the Civil War* (Princeton, NJ: Princeton University Press).
25. Michaels, Guy, Ashwani Natraj, and John Van Reenen (2014), "Has ICT Polarized Skill Demand?" *Review of Economics and Statistics* 96 (1): 60–77.
26. Condon, Bernard, and Paul Wiseman (2013), "Millions of Middle-Class Jobs Killed by Machines in Great Recession's Wake," *Huffington Post*, January 23, http://www.huffingtonpost.com/2013/01/23/middle-class-jobs-machines_n_2532639.html.
27. *The Economist* (2010), "Automatic Reaction," September 9, http://www.economist.com/node/16990700.
28. International Labour Organization (2015), *World Employment and Social Outlook Trends 2015* (Geneva: ILO).
29. Trading Economics (2015), "United States Youth Unemployment Rate," Last Accessed July 23, http://www.tradingeconomics.com/united-states/youth-unemployment-rate.
30. Boffey, Daniel (2015), "Youth Unemployment Rate Is Worst for 20 Years, Compared with Overall Figure," *The Guardian*, February 22, Last Accessed July 23, 2015, http://www.theguardian.com/society/2015/feb/22/youth-unemployment-jobless-figure.

31. Trading Economics. "France Youth Unemployment Rate," Last Accessed July 23, 2015, http://www.tradingeconomics.com/france/youth-unemployment-rate.
32. Greenspun, Philip (2009), "Technology Reduces the Value of Old People," *Philip Greenspun's Weblog*, October 29, Last Accessed September 14, 2015, http://blogs.law.harvard.edu/philg/2009/10/29/technology-reduces-the-value-of-old-people.
33. Kurtzleben, Danielle (2012), "Report: America Lost 2.7 Million Jobs to China in 10 Years," *US News & World Report*, August 24, Last Accessed September 14, 2015, http://www.usnews.com/news/articles/2012/08/24/report-america-lost-27-million-jobs-to-china-in-10-years.
34. Wright, Joshua (2012), "The Demise of Retail Jobs? Not So Fast," *emsi*, April 16, http://www.economicmodeling.com/2012/04/16/the-demise-of-retail-jobs-not-so-fast.
35. Bureau of Labor Statistics (2012), "Occupational Outlook Handbook," http://www.bls.gov/ooh/media-and-communication/reporters-correspondents-and-broadcast-news-analysts.htm.
36. Hilliard, John (2014), "US Jobs Grow, but Not for Journalists," *Christian Science Monitor*, March 7, Last Accessed September 14, 2015, http://www.csmonitor.com/Business/new-economy/2014/0307/US-jobs-grow-but-not-for-journalists.
37. Connelly, Brian L. et al. (2014), "Tournament Theory: Thirty Years of Contests and Competitions," *Journal of Management* 40 (1): 16–47.
38. Caves, Richard E. (2002), *Creative Industries: Contracts Between Art and Commerce* (Cambridge, MA: Harvard University Press).
39. MacDonald, Glenn M. (1988), "The Economics of Rising Stars," *The American Economic Review* 78 (1): 155–166.
40. DeVany, Arthur (2004), *Hollywood Economics: How Extreme Uncertainty Shapes the Film Industry* (New York: Routledge).
41. Aris, Annet and Jacques Bughin (2012), *Managing Media Companies: Harnessing Creative Value*, 2nd ed. (Hoboken, NJ: Wiley).
42. School of Information Management & Systems, University of California, Berkeley (2000), "How Much Information," Last Accessed May 14, 2008, http://www2.sims.berkeley.edu/research/projects/how-much-info/summary.html#consumption.
43. Picard, Robert G. (2004), "Environmental and Market Changes Driving Strategic Planning in Media Firms," In *Strategic Responses to Media Market Changes. Media Management and Transformation*, edited by Robert G. Picard (Jönköping, Sweden: Jönköping International Business School LTD).
44. Simon, Herbert (1971), "Designing Organizations for an Information-Rich World," In *Computers, Communication, and the Public Interest*, edited by Martin Greenberger (Baltimore: The Johns Hopkins Press), 37–72.
45. Collis, D.J., P.W. Bane, and S.P. Bradley (1997), "Winners and Losers–Industry Structure in the Converging World of Telecommunications,

Computing, and Entertainment," In *Competing in the Age of Digital Convergence*, edited by D.B. Yoffie (Boston: Harvard Business School Press).

46. Ibid.
47. Baumol, William and William Bowen (1966), *Performing Arts, the Economic Dilemma: A Study of Problems Common to Theater, Opera, Music, and Dance* (New York: Twentieth Century Fund).
48. Caves, Richard E. (2000), *Creative Industries: Contracts Between Art and Commerce* (Cambridge: Harvard University Press).
49. De Vany, Arthur (2004), *Hollywood Economics: How Extreme Uncertainty Shapes the Film Industry* (New York: Routledge), 231–254.
50. United States Department of Labor. "Futurework—Trends and Challenges for Work in the 21st Century Executive Summary," http://www.dol.gov/oasam/programs/history/herman/reports/futurework/execsum.htm.
51. Wu, Tim (2003), "Network Neutrality, Broadband Discrimination," *Journal of Telecommunications and High Technology Law* 2 (1): 141–176.
52. Raymond, Eric (2001), *The Cathedral and the Bazaar: Musings on Linux and Open Source by an Accidental Revolutionary* (Sebastopol, CA: O'Reilly Media).
53. Stallman, Richard (2007), "Reevaluating Copyright: The Public Must Prevail," *Oregon Law Review* 75 (Spring 1996): 291–297; Moglen, Eben. *Framing the Debate: Free Expression Versus Intellectual Property, the Next Fifty Years* (Barcelona: Universitat Oberta de Catalunya).
54. Rotenberg, Marc (2006), *Privacy & Human Rights* (Washington, DC: EPIC).
55. Noam, Eli, and Lorenzo Pupillo, eds. (2006), *Peer to Peer as a Distribution Medium* (New York: Springer, 2008); Benkler, Yochai. *The Wealth of Networks: How Social Production Transforms Markets and Freedom* (New Haven, CT: Yale University Press).
56. Von Hippel, Eric (2006), *Democratizing Innovation* (Cambridge, MA: MIT Press).
57. Noam, Eli (1998), "Spectrum Auction: Yesterday's Heresy, Today's Orthodoxy, Tomorrow's Anachronism. Taking the Next Step to Open Spectrum Access," *Journal of Law and Economics* 41 (2): 765–790.
58. Mossberger, Karen (2003), *Virtual Inequality: Beyond the Digital Divide* (Washington, DC: Georgetown University Press).
59. Lehr, William, and Lee McKnight (2003), "Wireless Internet access: 3G vs. WiFi?" *Telecommunications Policy* 27 (5–6): 351–370.
60. Smith, Aaron, and Janna Anderson (2014), "AI, Robotics, and the Future of Jobs," *Pew Research Center*, August 6, http://www.pewinternet.org/2014/08/06/future-of-jobs.
61. Atkinson, Robert D. (2013), "Stop Saying Robots Are Destroying Jobs— They Aren't," *MIT Technology Review*, September 3, Last Accessed April 24, 2017, http://www.technologyreview.com/view/519016/stop-saying-robots-are-destroying-jobs-they-arent/.

62. Markoff, John (2012), "Skilled Work, Without the Worker," *The New York Times*, August 18, http://www.nytimes.com/2012/08/19/business/new-wave-of-adept-robots-is-changing-global-industry.html.

References

Aris, Annet, and Jacques Bughin. 2012. *Managing Media Companies: Harnessing Creative Value*, 2nd ed. Hoboken, NJ: Wiley.

Asay, Matt. 2014. "The Internet of Things Will Need Millions of Developers by 2020." *Readwrite.* June 27. http://readwrite.com/2014/06/27/internet-of-things-developers-jobs-opportunity.

Atkinson, Robert D. 2013. "Stop Saying Robots Are Destroying Jobs—They Aren't." *MIT Technology Review.* September 3. http://www.technologyreview.com/view/519016/stop-saying-robots-are-destroying-jobs-they-arent/. Last Accessed April 24, 2017.

Atkinson, Robert D. et al. 2012. "Worse Than the Great Depression: What Experts Are Missing About American Manufacturing Decline." *The Information Technology & Innovation Foundation.* March. http://www2.itif.org/2012-american-manufacturing-decline.pdf. Last Accessed September 14, 2015.

Autor, David. 2010. "Polarization of Job Opportunities in the U.S. Labor Market." *Center for American Progress and the Hamilton Project.* April. https://www.brookings.edu/wp-content/uploads/2016/06/04_jobs_autor.pdf. Last Accessed April 24, 2017.

Baumol, William, and William Bowen. 1966. *Performing Arts, the Economic Dilemma: A Study of Problems Common to Theater, Opera, Music, and Dance.* New York: Twentieth Century Fund.

Benkler, Yochai. 2006. *The Wealth of Networks: How Social Production Transforms Markets and Freedom.* New Haven, CT: Yale University Press.

Bensinger, Greg. 2013. "Apps Are Creating New Jobs." *Wall Street Journal.* March 5. http://online.wsj.com/news/articles/SB10001424127887323864304578320861732248742.

Boffey, Daniel. 2015. "Youth Unemployment Rate Is Worst for 20 Years, Compared with Overall Figure." *The Guardian.* February 22. http://www.theguardian.com/society/2015/feb/22/youth-unemployment-jobless-figure. Last Accessed July 23, 2015.

Bort, Julie. 2014. "Bill Gates: People Don't Realize How Many Jobs Will Soon Be Replaced By Software Bots." *Business Insider.* March 13. http://www.businessinsider.com/bill-gates-bots-are-taking-away-jobs-2014-3.

Briefing Investor. 2012. "Industries Destroyed by the Internet—A Reflection." July 26. http://www.briefing.com/investor/our-view/ahead-of-the-curve/industries-destroyed-by-the-internet--a-reflection.htm.

Brynjolfsson, Erik, and Andrew McAfee. 2011. *Race Against the Machine*. Lexington, MA: Digital Frontier Press.

Bureau of Labor Statistics. "Occupational Outlook Handbook 2012." http://www.bls.gov/ooh/media-and-communication/reporters-correspondents-and-broadcast-news-analysts.htm.

Caves, Richard E. 2000. *Creative Industries: Contracts Between Art and Commerce*. Cambridge, MA: Harvard University Press.

Collis, David J., P. William Bane, and Stephen P. Bradley. 1997. "Winners and Losers—Industry Structure in the Converging World of Telecommunications, Computing, and Entertainment." In *Competing in the Age of Digital Convergence*, edited by David B. Yoffie. Boston: Harvard Business School Press.

Condon, Bernard, and Paul Wiseman. 2013. "Millions of Middle-Class Jobs Killed by Machines in Great Recession's Wake." *Huffington Post*. January 23. http://www.huffingtonpost.com/2013/01/23/middle-class-jobs-machines_n_2532639.html.

Connelly, Brian L. et al. 2014. "Tournament Theory: Thirty Years of Contests and Competitions." *Journal of Management* 40 (1): 16–47.

De Vany, Arthur. 2004. *Hollywood Economics: How Extreme Uncertainty Shapes the Film Industry*. New York: Routledge.

Du Rausas, Matthieu Pélissié et al. 2011. "Internet Matters: The Net's Sweeping Impact on Growth, Jobs, and Prosperity." *McKinsey Global Institute*. May. https://www.mckinsey.com/industries/high-tech/our-insights/internet-matters.

Goos, Maarten, Alan Manning, and Anna Salomons. 2009. "Job Polarization in Europe." *American Economic Review* 99 (2): 58–63.

Gordon, Robert J. 2016. *The Rise and Fall of American Growth: The U.S. Standard of Living Since the Civil War*. Princeton, NJ: Princeton University Press.

Greenspun, Philip. 2009. "Technology Reduces the Value of Old People." *Philip Greenspun's Weblog*. October 29. http://blogs.law.harvard.edu/philg/2009/10/29/technology-reduces-the-value-of-old-people. Last Accessed September 14, 2015.

Hilliard, John. 2014. "US Jobs Grow, but Not for Journalists." *Christian Science Monitor*. March 7. http://www.csmonitor.com/Business/new-economy/2014/0307/US-jobs-grow-but-not-for-journalists. Last Accessed September 14, 2015.

International Labour Organization. 2015. *World Employment and Social Outlook Trends 2015*. Geneva: ILO.

Kurtzleben, Danielle. 2012. "Report: America Lost 2.7 Million Jobs to China in 10 Years." *US News & World Report*. August 24. http://www.usnews.com/news/articles/2012/08/24/report-america-lost-27-million-jobs-to-china-in-10-years. Last Accessed September 14, 2015.

Lehr, William, and Lee McKnight. 2003. "Wireless Internet Access: 3G vs. WiFi?" *Telecommunications Policy* 27 (5–6): 351–370.

MacDonald, Glenn M. 1988. "The Economics of Rising Stars." *The American Economic Review* 78 (1): 155–166.

Markoff, John. 2012. "Skilled Work, Without the Worker." *The New York Times.* August 18. http://www.nytimes.com/2012/08/19/business/new-wave-of-adept-robots-is-changing-global-industry.html.

Michaels, Guy, Ashwani Natraj, and John Van Reenen. 2014. "Has ICT Polarized Skill Demand?" *Review of Economics and Statistics* 96 (1): 60–77.

Mintel. 2015. "Market Sizes." http://marketsizes.mintel.com. Last Accessed June 23.

Moglen, Eben. 2007. *Framing the Debate: Free Expression Versus Intellectual Property, the Next Fifty Years.* Barcelona: Universitat Oberta de Catalunya.

Mossberger, Karen. 2003. *Virtual Inequality: Beyond the Digital Divide.* Washington, DC: Georgetown University Press.

Noam, Eli. 1998. "Spectrum Auction: Yesterday's Heresy, Today's Orthodoxy, Tomorrow's Anachronism. Taking the Next Step to Open Spectrum Access." *Journal of Law and Economics* 41 (2): 765–790.

Noam, Eli, and Lorenzo Pupillo (eds.). 2008. *Peer to Peer as a Distribution Medium.* New York: Springer.

Picard, Robert G. 2004. "Environmental and Market Changes Driving Strategic Planning in Media Firms." In *Strategic Responses to Media Market Changes. Media Management and Transformation,* edited by Robert G. Picard. Jönköping, Sweden: Jönköping International Business School.

Quelch, John. 2009. "Quantifying the Economic Impact of the Internet." *HBS Working Knowledge.* August 17. http://hbswk.hbs.edu/item/6268.html.

Raymond, Eric. 2001. *The Cathedral and the Bazaar: Musings on Linux and Open Source by an Accidental Revolutionary.* Sebastopol, CA: O'Reilly Media.

Reilly, Jill. 2013. "Booming Internet Sales 'Will Close 5,000 High Street Stores and Cost 50,000 Jobs'." *Daily Mail.* January 1. http://www.dailymail.co.uk/news/article-2255677/Booming-Internet-sales-close-5-000-High-Street-stores-cost-50-000-jobs.html.

Rotenberg, Marc. 2006. *Privacy & Human Rights.* Washington DC: EPIC.

School of Information Management & Systems, University of California, Berkeley. 2000. "How Much Information." http://www2.sims.berkeley.edu/research/projects/how-much-info/summary.html#consumption. Last Accessed May 14, 2008.

Simon, Herbert. 1971. "Designing Organizations for an Information-Rich World." In *Computers, Communication, and the Public Interest,* edited by Martin Greenberger, 37–72. Baltimore: The Johns Hopkins Press.

Smith, Aaron, and Janna Anderson. 2014. "AI, Robotics, and the Future of Jobs." *Pew Research Center.* August 6. http://www.pewinternet.org/2014/08/06/future-of-jobs.

The Economist. 2010. "Automatic Reaction." September 9. http://www.economist.com/node/16990700.

Thibodeau, Patrick. 2009. "Study: Internet Economy Has Created 1.2M jobs". *Computerworld.* June 10. http://www.computerworld.com/article/2525229/internet/study--internet-economy-has-create-1-2m-jobs.html.

Timberg, Scott. 2013. "Jaron Lanier: The Internet Destroyed the Middle Class." *Salon.* May 12. http://www.salon.com/2013/05/12/jaron_lanier_the_internet_destroyed_the_middle_class. Last Accessed July 23, 2015.

Trading Economics. "France Youth Unemployment Rate." http://www.tradingeconomics.com/france/youth-unemployment-rate. Last Accessed July 23, 2015.

Trading Economics. "United States Youth Unemployment Rate." http://www.tradingeconomics.com/united-states/youth-unemployment-rate. Last Accessed July 23, 2015.

United States Department of Labor. "Futurework—Trends and Challenges for Work in the 21st Century Executive Summary." http://www.dol.gov/oasam/programs/history/herman/reports/futurework/execsum.htm.

VisionMobile. 2014. "IOT: Breaking Free from Internet and Things: How Communities and Data Will Shape the Future of IOT in Ways We Can't Imagine." *VisionMobile.* June. http://www.visionmobile.com/product/iot-breaking-free-internet-things/.

Von Hippel, Eric. 2006. *Democratizing Innovation.* Cambridge, MA: MIT Press.

Wildman, Steve, and Monica Herrero (eds.). 2015. *The Business of Media: Changes and Challenges.* Lisbon: Media XXI.

Wright, Joshua. 2012. "The Demise of Retail Jobs? Not So Fast." *emsi.* April 16. http://www.economicmodeling.com/2012/04/16/the-demise-of-retail-jobs-not-so-fast.

Wu, Tim. 2003. "Network Neutrality, Broadband Discrimination." *Journal of Telecommunications and High Technology Law* 2 (1): 141–176.

9

Job Losses and the Middle Class: Canada and the USA, and the Possible Role of ICT

Leonard Waverman

9.1 Introduction

Since 2000, three telecoms related issues have arisen in prominence for economists, as well as politicians—first, the productivity decline since at least the mid 2000s—where is the vaunted productivity impact of Information and Communications Technology (ICT)? Second, is an attempt to understand "polarization"—a decline in jobs and wages in the "middle" of the income distribution? Third, is the growth of ICT the cause of this polarization?

Since 2005 productivity growth in the west is not growing as fast as in the previous ten years. This poor performance has negatively impacted economic growth since *productivity* is one of the three sources of long-term economic growth, the other two being the *rate of growth of labor* (the rate of growth of population and the participation rate or the proportion employed equals the percentage employed) and the *growth rate of capital*. Of these three sources, productivity is the one most examined, discussed, and written about as it appears to be the one factor that countries can affect in the medium term. Thus, the preoccupation with productivity is because it appears to be something more controllable than long-run trends in population growth or capital accumulation.

A variant of this paper was published in Communications & Strategies, No. 100, 4th quarter 2016, pp. 165–180.

L. Waverman (*)
DeGroote School of Business, McMaster University, Hamilton, ON, Canada

© The Author(s) 2018
L. Pupillo et al. (eds.), *Digitized Labor*, https://doi.org/10.1007/978-3-319-78420-5_9

At the same time as the productivity slowdown, there appears to be a rise in income inequality across many western nations, particularly the USA and Israel. This rise in inequality was brought to prominence with Thomas Piketty's best-selling book, *Capital in the Twenty First Century*.[1]

The data, at least for the USA, the country most studied in the recent literature, suggest a decline in the returns to the middle-wage group, and this "decline of the middle class" has led to much discussion among academics and also at political levels. The role of ICT has taken on a potential major role in explaining both the productivity decline as well as "the fall of the middle class" but in two quite different ways. First, was not ICT supposed to be the third major Industrial Revolution and if so, why is its impact on productivity so fleeting?[2] And second, are ICT investments to blame for "job polarization" since ICT replaces routine jobs: bank tellers, middle managers even lawyers whose jobs can be done more cheaply by software programs?[3]

To shed some light on these issues some key concepts are introduced and analyzed below. As well, relative wage and income performance are compared between Canada and the USA and one measure of relative ICT performance between Canada and the USA is examined in some detail.

I then turn to the near future to assess whether ICT's role is over and done. The route of productivity performance is likely always uneven, and with the impact of the great recession still here, I consider that it is impossible to conclude that ICT's role is over. Indeed, I would expect ICT's role principally through smartphones to yield large productivity advances in the future.

9.2   Productivity and Polarization

Wage inequality and rising income inequality are not new topics of the last decade. Simon Kuznets won the Nobel Prize for his studies on income and wage inequality beginning in the 1940s. His 1954 Presidential address to the American Economic Association highlighted growing income *equality* accompanied by rapid increases in per capita income in the USA since the 1920s. In the mid- to late 1970s, income inequality began to rise in the USA and Piketty and Saez (2003) stated "a new industrial revolution has taken place, thereby leading to increasing inequality, and inequality will decline again at some point, as more and more workers benefit from the innovations." Thus, at least initially, the rise of the "digital" revolution was initially thought to increase income inequality but then to decrease it.

The iPhone hit the market only 10 years ago—June 2007, thus we are, in my view, too early in assuming that we know the long-term evolution of this digital revolution.

Technical change does not bring the same rewards to all. Originally economists (Solow 1956) modeled technical change as labor saving. Note crucially that this does not mean that permanent unemployment would result from labor-saving inventions. Quite to the contrary, labor-saving inventions make society better off since the demand for labor is an economy-wide macroeconomic phenomenon. Labor saving devices by making labor more productive increase wages and GNP. The number employed is determined by total demand, exchange rates, exports, and imports, etc. In the shorter run, there are certainly dislocations for that labor that is displaced. Consider agricultural advances in productivity displacing agricultural workers who then migrated to towns in search of employment. We as a society and we as academic economists have done a very poor job in considering the plight of employees severely affected by technological shifts such as the growth of ICT's. We have, as a society, not offered sufficient retraining, guidance, and counseling and the social consequences are clear.

A new hypothesis emerged in the 1990s—that technical change was "skills biased" not just labor saving (Skills Based Technical Change—SBTC) and this change was a result of the rise of microprocessors in the 1980s (see Johnson 1997). David Card and John DiNardo (2002) showed however that SBTC did not explain the patterns of wage inequality of the 1980s and 1990s and that the SBTC hypothesis was inconsistent with other labor market facts such as the returns to education.

In 2003 yet another hypothesis was advanced by economists—that ICT enables the elimination of jobs which are routine based (RBTC). Thus the hypothesis became that middle-class jobs were disappearing because middle-class job functions (bank tellers as an example) were repetitive routine type jobs while high skilled jobs were more "cognitive" and low skilled service jobs were nonroutine. For example, one cannot at this point replace low skilled hospital orderly jobs with ICT. Thus advances in ICT lowered the demand for routine workers, many of whom were in the middle of the income distribution (see Fig. 9.1).

Goos and Manning (2007) show the changes in employment shares by occupation group in the USA for three time periods (1981–1991, 1991–2001, and 2001–2011) and for the three divisions of occupations discussed above—nonroutine manual (low skilled), nonroutine cognitive (high skilled) and the "middle" (routine). In their analysis, the polarization hypothesis appears to be borne out for the USA—routine type jobs have been

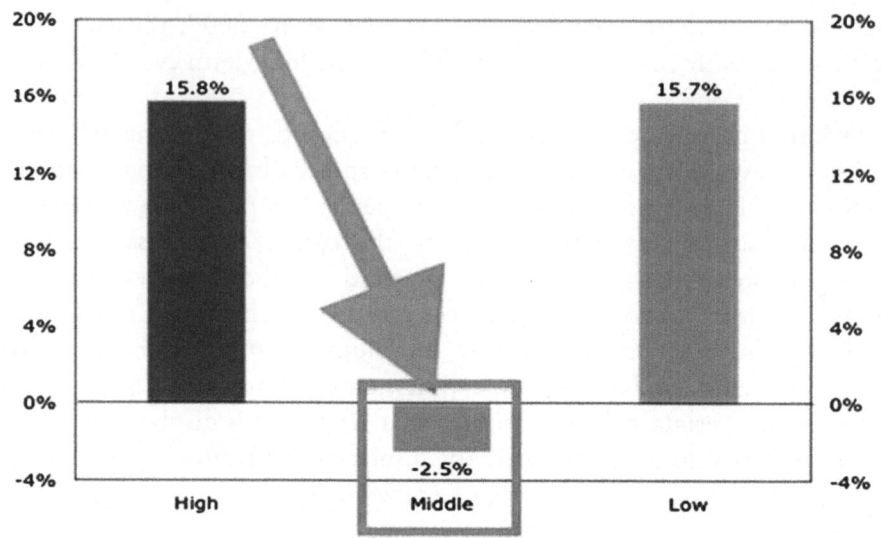

Fig. 9.1 Job growth and decline by skill level in the USA from 2003 to 2013 (*Source* Wells Fargo 2014)

disappearing and at an accelerating rate for three decades. Of course, many other changes have been occurring in the US economy as well over these 30 years, some but not all of these are perhaps affected by the rise in ICT's. One such change has been the rise of outsourcing; a second is the growth of Chinese manufacturing; a third is a move to more of a service-based economy; and a fourth (there are other changes as well) is the new trade agreements (NAFTA, for example).

The rise of China and Chinese manufacturing was based on low skilled Chinese labor (routine biased) migrating to cities from rural areas, like the previous industrial revolutions in England. ICT and global supply chains enabled some of this rise, and the Chinese labor supply replaced routine jobs in the west. But clearly to say that Chinese success is due to ICT (and no one does) is wrong. Nor are new trade agreements "due" to ICT advances. Nor is the rise of the service sector "due" to the rise of ICT; the percentage spent on services rises as incomes grow. Thus, we must be very careful in assessing the role of ICT in labor market changes—a complete general model is needed. It is in my opinion incorrect to make employment analyses at the overall macroeconomic level and to try to assess whether a characterization of jobs is the source of job losses without accounting for the many dynamics and major shifts in the economy. I also doubt that we can credibly assess job and occupational categories as routine, nonroutine, cognitive, etc. The studies cited above are undertaken at the economy-wide level, and assessing jobs/occupations even at 3 or 4 digit levels omits firm-specific details which are crucial. For example, in the auto sector in the 1980s, assembly line jobs may have been "routine" at many firms but clearly at Toyota they were "routine cognitive."

These caveats are substantial but still the economy wide data do show clear trends in polarization of wage, income, and job classifications especially for the USA. I turn to US–Canada comparisons as Canada is at the same development level as the USA and the two countries share the world's largest open border, language, and culture. Do the two countries share the same employment shifts?

9.3 Canada and the USA

9.3.1 Productivity and ICT

Canada and the USA have many similarities in their economies; many industries even have the same players. Yet Canada has a profile different from the USA in ICT use and in ICT proliferation.

Waverman and Dasgupta (2011) developed "The Connectivity Scorecard" (CS) concept. This scorecard measured and ranked a country's combination of communications infrastructure, usage of this infrastructure, skills and measures of business adaptability of advanced web, and ICT applications and services. For 25 advanced economies, CS utilized 25 different attributes for the three major GNP components: consumers, businesses, and government. Unlike most qualitative scorecards, CS used well-defined weights which were country-specific. Weights were drawn from the economics literature as well as for individual country GNP shares of consumption, business transactions, and government spending. The country that did "best" in any single component received a score of 10 for that component and all other countries were scored relative to that country. Countries were then ranked on their aggregate index, the maximum was 10. In 2011, the last time Waverman and Dasgupta authored the Connectivity Scorecard, Sweden ranked first, the USA second, and Canada eighth.

Below are Venn diagrams from the Connectivity Scorecard calculations for 2011 for the USA and Canada. These diagrams show the three sectors, business, consumers, and government for each country as well as the two components: infrastructure, and usage and skills scores.

Observing the Venn diagram for the USA (9.2a), the USA is the leader among all countries in two categories—consumer usage and skills and in business infrastructure (as the USA position is the farthest out). However, the USA lags the best performing country primarily in two areas—business usage and skills (mainly due to a fall in higher education in STEM areas) as well as in consumer infrastructure (at that time, a lag in broadband relative to the world leaders Japan and Korea).

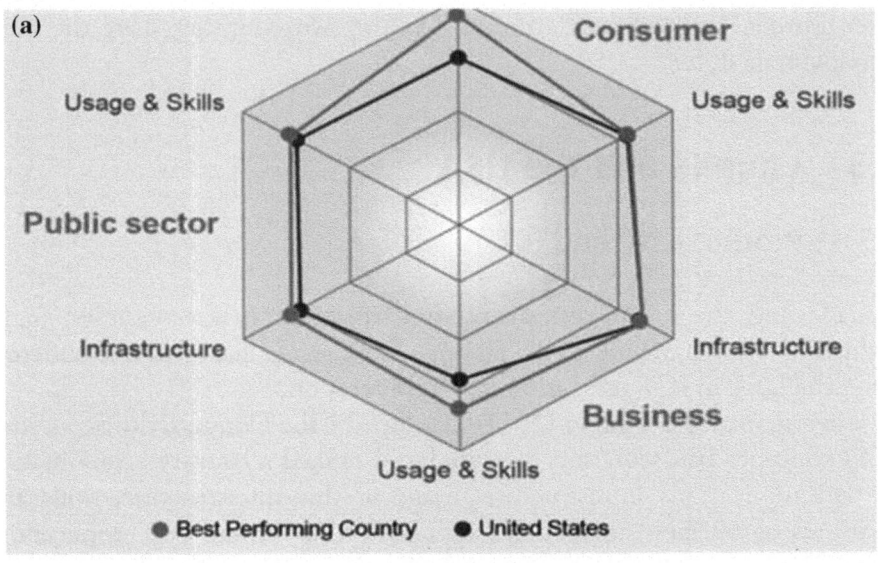

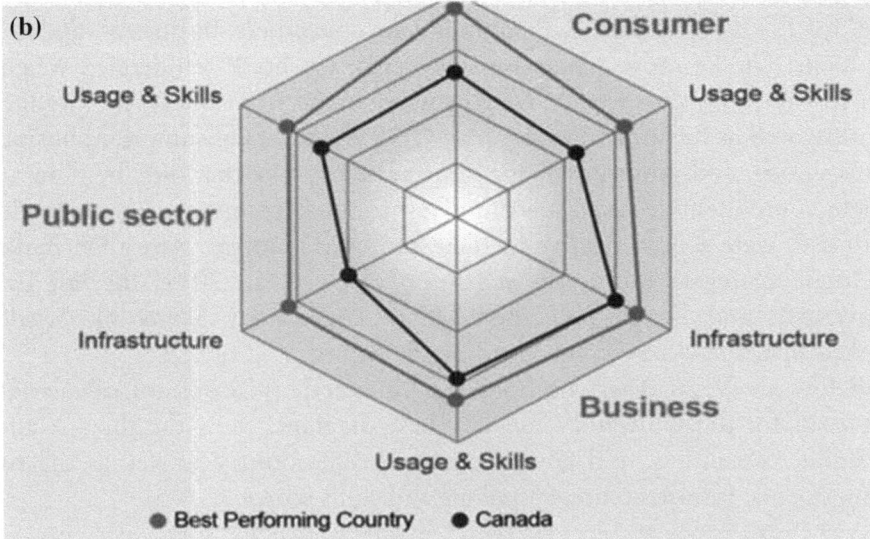

Fig. 9.2 Venn diagrams, connectivity scorecard 2011, USA and Canada

Turning to Fig. (9.2b), the Venn diagram for Canada, the differences with the USA (9.2a) are clear. Even though the two economies have similar styles of business and government, and have the largest bilateral trade in the world, ICT adaption, usage, and skills vary markedly between the two countries.[4] The Venn diagram for Canada is in effect inside that of the USA for five of the six components. Only the category "business usage and skills" is similar for the USA and Canada and the scores for both countries are also among the highest in the world. The differences between Canada and the USA are especially marked for consumer and government infrastructure and for business infrastructure.

Turning to productivity, Canadian productivity growth has consistently lagged that of the USA, as has the contribution to productivity from ICT. Figure 9.3 derived from Fuss and Waverman (2005) disaggregates the 2003 twenty-one percent productivity gap of that year between the USA and Canada into its component sources. We choose 2003 as that was a year when productivity performance in both economies was high and it is also a year when most researchers agree that ICT was a major cause of productivity growth in the USA and elsewhere.

Figure 9.3 is interpreted in the following way. Non-ICT capital differences between Canada and the USA account for only 5% of the 21 point difference between Canadian and USA productivity. Differences in the scale of the two economies (the USA is a much larger country) accounts for 15% of the 21 point productivity difference. Significantly, the lower ICT level in Canada relative to that in the USA accounts for over half of the productivity difference between the two countries!

At the right of Fig. 9.3, the components of this lower level of ICT in Canada are disaggregated. Of the 56% difference explained, only 12% is due to the lower ICT capital stock itself. The majority or 44% is due to what we label the *ICT spillovers or the characteristics of ICT*. A lower level of

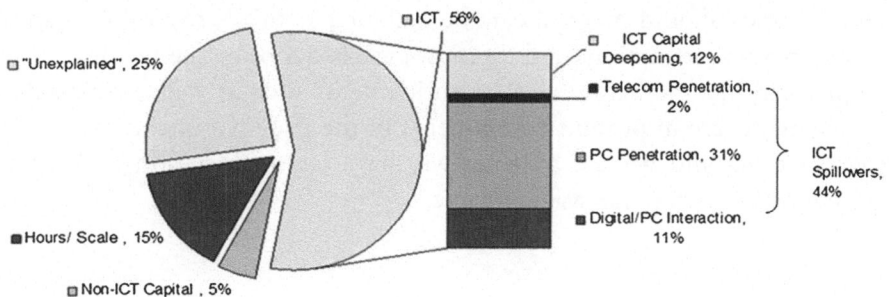

Fig. 9.3 Contributions to US–Canada productivity gap, 2003 (*Source* Fuss and Waverman 2005)

PC penetration in Canada accounts for 30 percentage points of this 56 percentage point difference.

Thus, two very similar economies, geographically next door to each other, with many US subsidiaries being major operating firms in Canada have very different ICT characteristics and performance. Productivity differences are large, ICT levels, usage, and skills differ and explain over half of the productivity differences between the two economies. And these differences persist for decades.

Explaining Canadian poor productivity performance and the role of ICT in that performance is a focus of research in Canada, yet it remains a puzzle.

The polarization of wages/incomes, the major research focus in labor economics in the USA over the past decade is now examined. As noted, aggregate US data show the loss of middle-class/routine jobs. Do Canadian data on employment demonstrate a similar polarization story?

9.4 Income/Wage/Job Polarization: USA and Canada

A number of recent analyses empirically examine the issues of polarization in Canada. I rely on one particular paper here as it examines both USA and Canadian data (Green and Sand 2013). As stated above, comparisons of wage disparities by income class or occupational grouping between two countries are not easy to make because of a variety of issues. There is also a lack of comparability of data. Issues of institutional governance also affect comparisons, as I now show.

Canada has a more progressive tax and welfare system than does the USA; Canada has a well-functioning universal health-care system; the US health-care system is privately funded except for the new provisions of Obamacare. Should we be examining before tax income or after tax and after entitlement income? Should medical care be included as this is funded in Canada through payroll deductions and taxes? So, the issues of before tax or after tax income, and income before or after entitlements such as welfare or medical care payments are important components in the issue. Countries with more progressive tax and welfare schemes will have less polarization in after tax income than in before tax wage income.

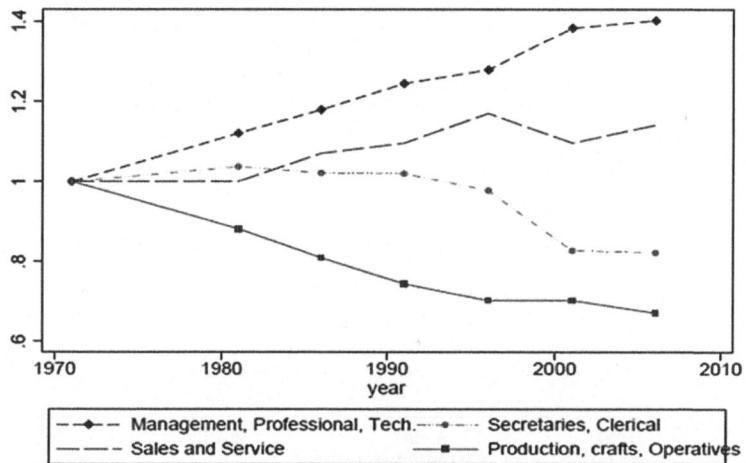

Data comes from the Canadian Census Master Files from 1971-2006. The figure represents the share of hours worked among four broad occupation classifications, indexed to 1 in 1971.

Fig. 9.4 Canada: Share of hours worked 1970–2010 (*Source* Green and Sand 2013)

We begin with data on employment. Figure 9.4 presents a general picture of employment trends across occupations in Canada from Green and Sands (2013). The data show Canadian employment distribution (hours worked) among four classifications: management, professional, technical; sales and service; secretarial, clerical; production, crafts and operatives for the 1970–2010 period. The latter two job classifications are more likely to be routine-based jobs. Normalizing at 1 for all four job categories in 1971, one can see relative growth in the first category, managerial, professional, and technical. For the second category, sales and service employment, its share of all employment rose from 1980 but falls from 1995 to 2000 and then increases again.

The crafts and operatives category share has been in decline since 1971 with a leveling out in the 1995–2000 period and a decline again post 2000. The category secretaries and clerical share of employment rose slightly to 1980 fell slightly to 1990 then fell to 2000 especially in the 1995–2000 period.

The US data in Fig. 9.5 show similar but more pronounced movements particularly in the post-2000 period.

Green and Sand (2013) summarize their results for Canada for this period as follows:

We find that there has been faster growth in employment in both high and low paying occupations than those in the middle since 1981. However, up to 2005, the wage pattern rejects a simple increase in inequality with greater growth in high

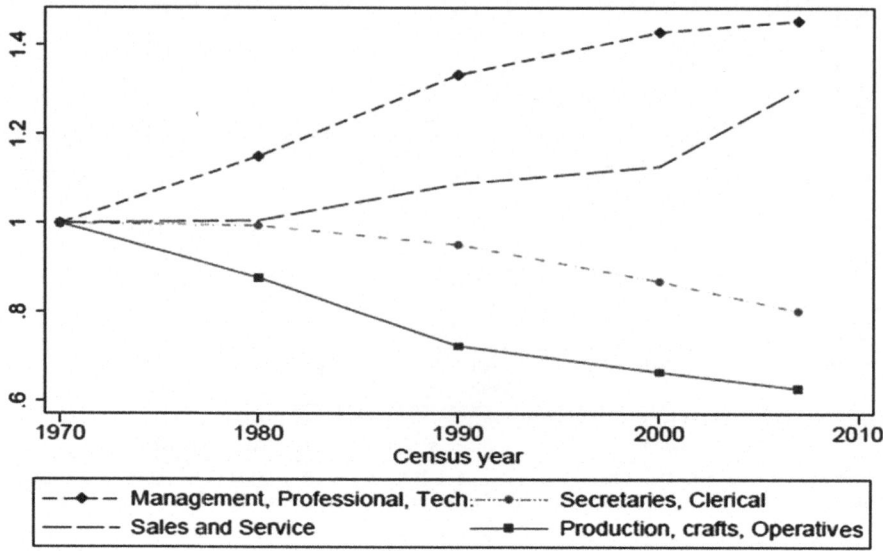

Fig. 9.5 USA: Share of hours worked 1970–2010 (*Source* Green and Sand 2013)

*paid than middle paid occupations and greater growth in middle than low paid occupations. Since 2005, there has been some polarization but this is present only in some parts of the country and seems to be related more to the resource boom than technological change. We present results for the US to provide a benchmark. The Canadian patterns fit with those in the US and other countries apart from the 1990s when the US undergoes wage polarization not seen elsewhere. **We argue that the Canadian data do not fit with the standard technological change model of polarization developed for the U.S.** (emphasis added)*

Green and Sand also state:

In a study that compares movements in both employment and wages between the U.S. and Germany, Antonczyk, DeLeire, and Fitzenberger (2010) find that, although there are similarities in occupational employment between the two countries that is consistent with technological change, the differences in the evolution of the wage distribution between the two countries is so large that technology alone cannot explain the wage trends.

We turn to another comparison for the USA and Canada, examining wage movements rather than occupational shares. Figures 9.6 and 9.7 reproduce data from Green and Sand (2013) on the percentage change in weekly wages (e.g., change in weekly wages by wage percentile 1991–2001) for Canadian men (Fig. 9.6) and for US men (Fig. 9.7). The "hollowing out" hypothesis is that wage changes in the middle of the distribution are negative.

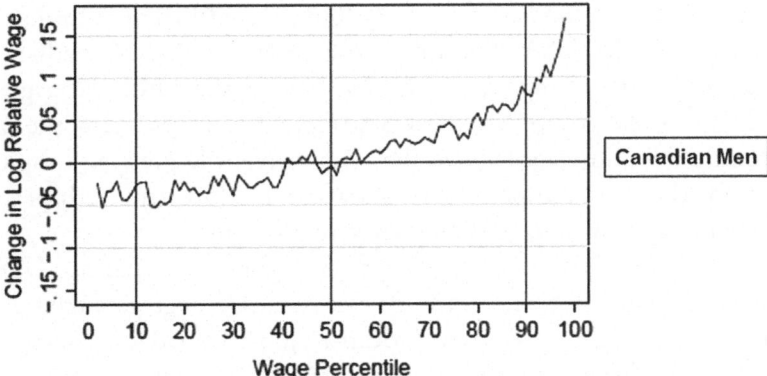

Fig. 9.6 Change in log weekly wages by percentile change from 1991 to 2001 (*Source* Green and Sand 2013, pp. 12 and 13)

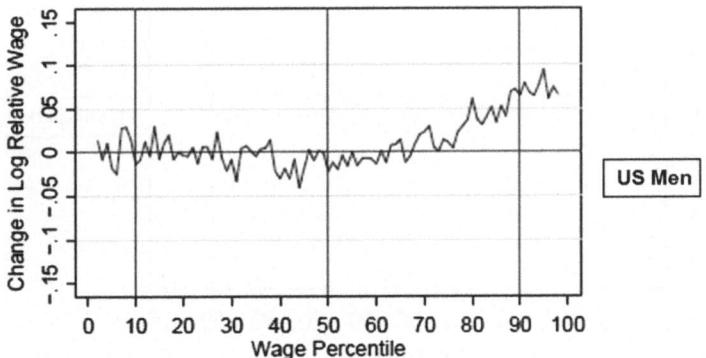

Fig. 9.7 Change in log weekly wages by percentile change from 1990 to 2000 (*Source* Green and Sand 2013, pp. 12 and 13)

For Canada (Fig. 9.6), in the 1990s, to summarize, median wages show a near-ubiquitous increase in wage inequality over the entire range. There are wage decreases below the 40th wage percentile and increases above that level, *but there appears to be no "polarization"*—wages in the middle segment did not fall relative to higher or lower wages.

The US wage data pattern in the 1990s (see Fig. 9.7) is markedly different from these Canadian data (and from most European patterns as well). In the USA, the pattern is not linear, but there are modest wage increases up to the 30th wage percentile, modest decreases to the 70th percentile (polarization?) and increases thereafter.

For the period 2001–2006, the wage patterns are as follows for Canada and the USA:

When we examine the data for 2001–2006, very different patterns emerge for Canada and the USA. For this more recent period, Canadian wages (Fig. 9.8) show little change up to the 70th percentile while wages grew 5% for the higher wage group. For the USA (Fig. 9.9), wages below the 25th or so percentile fell and wages above the 55th percentile grew and rapidly, by 10–15% for the top 90th percentile. Between the 25th and 55th percentile, US wage rates appear to have been relatively stagnant.

To summarize this brief survey of one paper comparing USA and Canadian occupational share and wage rate data, US aggregate data do show job classification shifts away from "routine" jobs; Canadian data appear to not show such shifts. Examining percentage changes in wage data, some hollowing out of the "middle class" is evident in the USA. No such pattern

Fig. 9.8 Change in log weekly wages by percentile change from 2001 to 2006

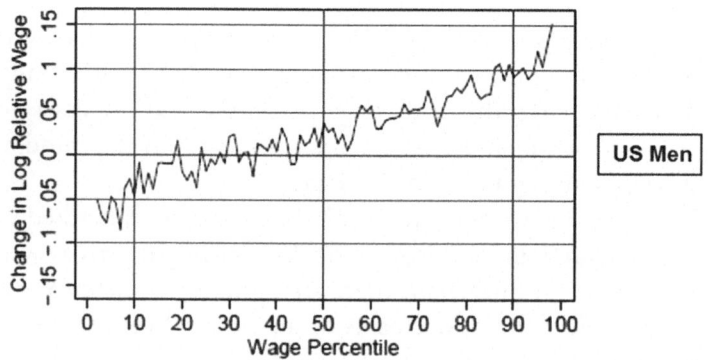

Fig. 9.9 Change in log weekly wages by percentile change from 2000 to 2007 (*Source* Green and Sand 2013)

is seen in Canada. And these data are for pretax percentile wage distribution data. On an after tax basis, wage polarization would not appear to typify aggregate Canadian data 1991–2006.

9.5   Relation to ICT?

We saw earlier that Canada lags the USA in many types of ICT infrastructure, adaption, applications, and usage. The largest gaps in 2011 were the lower levels in Canada in ICT business, government and consumer infrastructure, as well as in Canadian consumer usage of ICT.

We saw earlier that business ICT usage and skills are very similar in Canada and the USA. So, the poorer Canadian productivity performance cannot be because of this. And it is hard to see how the differences that do exist between Canada and the USA in ICT—in consumer and government infrastructure could have large productivity impacts as productivity is largely a business phenomenon. The data in Fuss and Waverman (2005) do show a significant lower adaption of computers in Canada—these differences are suspicious and could be due to data errors but if these are true differences, this one ICT capital stock difference would be a significant factor.

Of course, the widely reported and discussed fall in worldwide productivity since 2007, particularly in the USA is both perplexing and troubling. Figure 9.10 shows the productivity experience in the USA since 1947. Until 1972, productivity growth in the USA was at very high levels, some 2.8% per year. As Robert Gordon stresses, this postwar period saw many technological advances as well as the postwar recovery period. From 1973 through 1995, productivity growth in the USA (and in most of the world) was tepid, averaging just above 1.5% per year. Policy makers, analysts, and economists bemoaned the stagnation of productivity growth over this 20 year period. The period 1995–2007 is clearly well above the productivity performance of the period from 1973 to 1990. Most researchers cited ICT developments as the reason for this growth spurt.

> ... the underlying cause was an increase in the rate of decrease of semiconductor prices and, in turn, of ICT capital equipment. In response to falling ICT prices, producers in both services producing and goods-producing sectors shifted increasing amounts of capital investment toward ICT products, reducing in some cases purchases of more traditional capital equipment. Subsequently, many business analysts have noted that, following a gestation lag, the lower cost of ICT equipment has induced firms to "make everything digital" and reorganize their business practices (Anderson and Kliesen 2006)

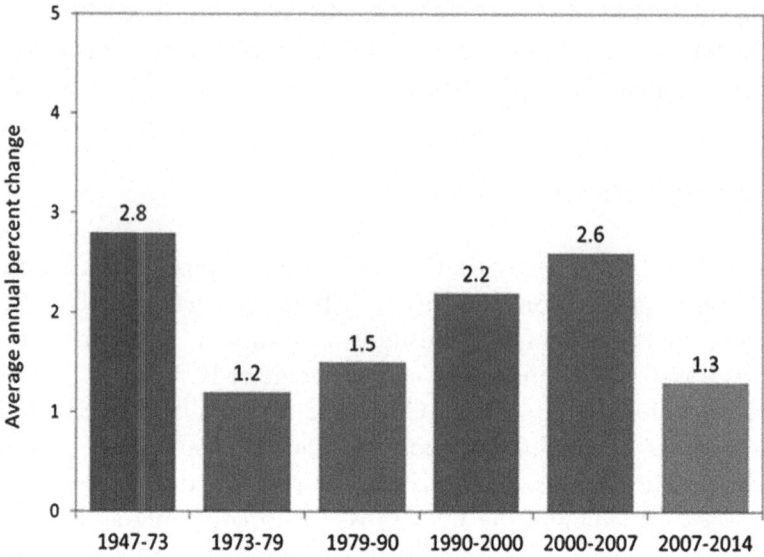

Fig. 9.10 Productivity change in the US nonfarm business sector, 1947–2014 (*Source* US Bureau of Labor Statistics)

Note the fall off in productivity growth since 2007, through 2014, rising at only 1.3% per year—are we back to the future—the relatively low growth of 1973–1995, is that the new normal? This is what Robert Gordon argues in his new book (2016). Professor Gordon argues that ICT developments are not of the same order of magnitude, longevity or "general purpose" as steam or the internal combustion engine or other products that have had long-term significant productivity boosting impacts.

However, I think that "it's too early to tell," that we as yet do not know whether we are back to the tepid productivity growth of 1973–1995.

Remember that the World Wide Web dates to 1996, 21 years ago. We are thus not far down the ICT path. Nor can we expect monotonic improvements in welfare and productivity given economic shifts unrelated to ICT such as the Great Recession of 2008, the dislocations of which are still being felt worldwide.

The explosion of social media, viral networks and applications spawned in 2007 by the iPhone are now but ten years old! Most advances have been directed at the consumer market. Even there, new advances such as self-driving cars and virtual reality are not market ready. *We have not yet begun to tap the enormous business potential of the ubiquitous smart phone.* It is critical to remember that other general purpose inventions were slow to come to fruition and uneven in timing and impact. "…steam had a relatively small and long-delayed impact on productivity growth…" (Crafts 2002). In a well-cited paper, Paul David (1989) states "…the transformation of industrial processes by the new electric power technology was a long delayed and far from automatic business."

Thus my conclusion, "it's too early to tell."

Productivity will improve as the ICT revolution continues to expand beyond consumer-driven social media. Certainly the US data do appear to show that ICT is, at least, in the short to medium term a source of growing income disparity because of the displacement of "routine" jobs. In the steam era, the displacement of agricultural workers due to mechanization in agriculture increased the labor force needed for the new factories. Today, the displacement of routine noncognitive jobs by ICT has not led to a parallel expansion of employment in jobs spawned elsewhere by ICT. That is, the adjustments for routine labor jobs in earlier technological shifts were in essence self-reinforcing. The farmer was displaced, costly moves were required to migrate vast distances to jobs in urban metropolises. No one wants to minimize these costs. However, today the skill levels required are to move from routine-based jobs to nonroutine or cognitive occupations. These are very different transitions from the past. Rich countries—the West—have done far too little to enable such job/occupational shifts. And for those who cannot at their life cycle stage make such adjustments, we have not created the needed social safety nets. All countries need to look at the US experience. And we need to prepare for these coming changes by identifying at risk groups and preparing counseling, training and income relief far better than we have in the past.

Notes

1. The potential measures of income inequality used in the literature are many, and can be measured as inequality of total income, inequality of income with or without capital gains and dividends, inequality of wage income, inequality of wealth, etc. The ways to measure income (or wage) inequality are also many, and include: the percentage of total country income earned by the top x%, usually the top 10% or 1%; the Gini Coefficient measuring deviations from equal income distribution; and now the "polarization" effect: comparing wage changes over time for certain classes of wage earners, generally "low," "middle" (the middle class), and "high."

2. Gordon, Robert (2016), *The Rise and Fall of American Growth* (Princeton: Princeton University Press).

3. Note that these two concerns about ICT are in essence mutually exclusive since if ICT does remove many jobs, then productivity (which is measured as output change minus capital and labor force changes) should be growing.
4. See Alan Blinder on the productivity slowdown in the USA, Wall Street Journal May 14, 2015. Some authors pick 2005 as the date of the slowdown, some 2007 and some 2010! In part 5 I show data for pre- and post-2007 productivity growth.

Bibliography

Anderson, Ricahrd G., and Kevin L. Kliesen. 2006. The 1990s Acceleration in Labor Productivity: Causes and Measurement, February 2006: 2, 3. Research Division, Federal Reserve Bank of St. Louis, mimeo.

Antonczyk, Dirk, Thomas DeLeire, and Bernd Fitzenberger. 2010. *Polarization and Rising Wage Inequality: Comparing the US and Germany.* Germany: Institute for the Study of Labour, IZA DP No. 4842, mimeo.

Card, David, and John E. DiNardo. 2002. "Skill Biased Technological Change and Rising Wage Inequality: Some Problems and Puzzles." *Journal of Economic Literature* 20 (4): 733–783.

Crafts, Nicholas. 2002. *Productivity Growth in the Industrial Revolution: A New Growth Accounting Perspective.* London: London School of Economics, mimeo.

David, Paul. 1989. The Dynamo and the Computer: An Historical Perspective on the Modern Productivity Paradox. Nashville: American Economic Association, Papers and Proceedings.

Fuss, Melvyn, and Leonard Waverman. 2005. *Canada's Productivity Dilemma: The Role of Computers and Telecom.* Canada: Bell Canada's Submission to the Telecommunications Policy Review Panel, Appendix E-1.

Goos, Maarten, and Alan Manning. 2007. "Lousy and Lovely Jobs: The Rising Polarization of Work in Britain." *The Review of Economics and Statistics* 89 (1 February): 118–133.

Green, David A., and Benjamin Sand. 2013. *Has the Canadian Labour Market Polarized?* Vancouver: Vancouver School of Economics, University of British Columbia, mimeo. November. Revised Version Published in the *Canadian Journal of Economics*, 48 (2), 2015. The figures in this paper are taken from the 2013 paper.

Johnson, George E. 1997. "Changes in Earnings Inequality: The Role of Demand Shifts." *Journal of Economic Perspectives* 11 (Spring): 41–54.

Kuznets, Simon. 1955. "Presidential Address American Economic Association." *American Economic Review* XLV (1): 1.

Piketty, Thomas, and Emmanuel Saez. 2003. Income Inequality in the United States 1913–1998. *Quarterly Journal of Economics* CXVIII (1) (February).

Piketty, Thomas. 2014. *Capital in the Twenty-First Century*. Cambridge: Harvard University Press.

Solow, Robert. 1956. "A Contribution to the Theory of Economic Growth." *Quarterly Journal of Economics* 70 (1): 65–94.

Waverman, Leonard, and Kalyan Dasgupta. 2011. *The Connectivity Scorecard*. Finland: Nokia Siemens Networks, mimeo.

Wells Fargo. 2014. *The Face of Job Polarization*. Special Commentary September 24, 2014, written by E. Aleman and A. Khan. Note That These Data May Have Been Revised from What Was Published Then by the U.S. Department of Labor.

10

Internet Innovations–Software *Is* Eating the World: Software-Defined Ecosystems and the Related Innovations Result in a Programmable Enterprise

Robert B. Cohen

10.1 The Internet as a Driver for Innovations

The Internet has been the foundation for innovative changes in the use of computing, networking, and storage over the past decade. One major innovation, software-defined ecosystems,[1] has provided a platform for a series of innovations that would only have occurred if today's dramatically altered infrastructure—cloud computing infrastructure—was available to support them. These innovations have transformed how businesses operate, how software is developed, how we analyze data.

What drives these Internet innovations is a major structural change from the "Old IP" to the "New IP." The Old IP was the dominant Internet infrastructure from the early 1990s until recently. The "Old IP" is hardware-centric, with a rigid topology and architecture, innovation depended on proprietary but standards-based innovation. By contrast, the "New IP" is software-centric, fluid in topology and architecture, and capable of scaling clients and resources on demand.[2]

A basic innovation, software-defined infrastructure, lies at the heart of the "New IP," software-defined ecosystems. Facebook, Amazon, Netflix, and Alphabet (formerly Google) have built their operations on these ecosystems. The infrastructure they have built lets them benefit from the enormous growth of mobile video and data traffic. These firms have also commercialized business operations that underpin a vast expansion of social media and data analytics.

R. B. Cohen (✉)
Economic Strategy Institute, Washington, DC, USA

© The Author(s) 2018
L. Pupillo et al. (eds.), *Digitized Labor*, https://doi.org/10.1007/978-3-319-78420-5_10

10.2 Software-Defined Infrastructure: The Central Software Innovation

The Internet's key innovation is software-defined infrastructure. This innovation is the cornerstone of software-defined environments. In these environments, software not only replaces hardware (creating software-defined infrastructure) at the heart of computing, storage, and communications networks, but also becomes the platform and prime tool to develop new business processes, create new applications, and perform more complex data analysis.

This essay finds that contrary to strongly held views that the Internet's contribution to innovation, productivity, and growth ended years ago,[3] the adoption of software-defined environments has expanded the products, services, and markets that the economy supports. Thus, the expression, "software will eat the world,"[4] namely, that software will not only disrupt traditional industries, but also add significant efficiencies to existing industries, is not ill-considered. Through software-based innovations, the Internet will expand our economic base far beyond its current limits. It will promote job growth and improve productivity.

This will occur as innovations linked to software-defined environments change the way infrastructure functions. We begin by describing software-defined environments. We then explore how this fundamental innovation alters not only how we develop software, but also manage and analyze data and complex networks. These are the knock-on innovations that are tied to the core change, software-defined environments. These secondary innovations are central to recent advances in the economy. Without them businesses would find it impossible to manage production, optimize supply chains, or to operate real-time sensor networks for aircraft management, driverless vehicles, agricultural planting and harvesting, and wearable devices.

10.3 Software-Defined Environments: Controlling Today's Infrastructure

Software-defined environments are based upon software-defined infrastructure (SDI). This infrastructure relies upon software controls and instructions. It distinguishes today's infrastructure from that at the beginning of the Internet Age (1990–2000). SDI is a new "software platform." As such, it manages operational requirements that hardware once defined, i.e., software replaces the functions of physical equipment, such as servers, switches, load balancers, and intrusion detection devices.

When infrastructure shifts to being software-defined, software commands direct infrastructure to perform in a way that is more flexible, agile, scalable, and interoperable than was possible 15 years ago. If a group needs to test new software by scaling to a "business-wide operational level" involving access to tens of thousands of servers, this can be done in seconds because the software-defined environment can amass the required servers inside an enterprise or from the public cloud—from providers, such as Amazon AWS or Microsoft Azure—to do the testing in seconds instead of weeks or months.

10.4 The Move to OpenStack and APIs

The emergence of OpenStack and APIs has made it much easier for companies to transform their infrastructure and adopt software-defined ecosystems.

OpenStack is possible because of the open source model that many firms have decided to adopt. The open source movement began with the free release of the code for Netscape,[5] an early Internet browser over 25 years ago. Since open source software is free, it changes the economics of building software-defined infrastructure. Many firms, including Netflix, place their software on an Internet-based exchange named GitHub, where it is licensed for free use by others. AT&T has said it will increase the amount of open source software that it is using from 5% in 2015 to about 50% by 2020.[6]

Open source software encompasses a broader movement of technologists that is called OpenStack. OpenStack is "a free and open-source software platform for cloud computing …. The software platform consists of interrelated components that control hardware pools of processing, storage, and networking resources throughout a data center. Users either manage it through a web-based dashboard, through command-line tools, or through a RESTful API."[7]

APIs or Application Programming Interfaces let programmers interact with private and public OpenStack clouds. APIs are a set of techniques used "by computer programs to request services from the operating system, software libraries or any other service providers [are] running on the computer." By using APIs, developers can provide access to a proprietary software application. An API's code includes "requirements that govern how one application can communicate and interact with another. They also allow developers to access certain internal functions of a program."[8] More simply, APIs permit one software to interact with another. This lets developers in one firm create interfaces that can access software or developers in another organization to access an interface by using APIs.[9]

The real value of APIs is to make it possible for people and/or software—"automated service discovery"—to identify ways to manage other software and to automate business processes. While this is expected to happen in the future, the benefits of automating business processes are likely to be great. Thus, APIs might come to the fore more rapidly than expected.[10]

10.5 The Emergence of Continuous Software Delivery and DevOps: Creating Software Using Software

Agile software-defined environments have transformed how businesses design software. One innovation, "Continuous Service Delivery"[11] changes software development from a highly siloed series of processes—where every part in the software creation process took days to complete—to a new collaborative team structure. DevOps[12] is a complementary process change that accelerates software development by making processes more agile and lean. It also provides the basis for automating segments of the software development process, initially by using modular subparts of software or "microservices." Recent surveys show that many firms have already optimized this process and are able to deploy a new application or service in less than 15 minutes.[13]

10.6 Containers: An Innovation to Deploy and Manage Software and Data

A related innovation is containerization. Containers are a software-defined structure that is attached to servers that define applications and the environment in which they will operate, as well as the data they can use. This innovation has simplified the software development process so greatly that some firms are running hundreds of containers on a single server. With containers, resources behave as though they are part of a single, unified environment.

The main benefit of containers is to accelerate software implementation at different locations. Containers also create a pipeline for the distribution of new services. In addition, containers support building new software once and running it on different platforms.

10.7 Federating Data and Providing for Data Collection at the Edge of Networks ("Fog Computing")

Many firms are analyzing large databases. In the case of firms in several major industries, this has only been possible with the consolidation of data centers. So, Boeing, Ford, GM, UPS, and FedEx, as well as Netflix, Facebook, and other firms have created "big data" lakes they analyze by using sophisticated analytic tools.

For aerospace and auto firms, the analysis of "big data" streamlines the operations of suppliers around the globe. In other industries, firms, such as health-care providers are linking the analysis of "big data" genomic information to treatment outcomes. By doing so, they are transforming health care.

"Fog Computing" is the decentralization of infrastructure to support driverless cars and wearable devices. Today's software-defined ecosystems support this decentralization, where computing, storage, and networking, to support sensor networks, are relocated to the periphery of networks. This changes the architecture of infrastructure by relocating equipment and software to end points in data networks. This restructuring lets such peripheral centers collect data from planes, cars, and wearers of health devices collected transmissions close to their source and provides a way to evaluate them quickly.

10.8 Block Chains and More Secure Infrastructure

Block Chains[14] are "a public transaction ledger built in a network structure based on cryptographic principles so there does not need to be a centralized intermediary. Any kind of asset (art, car, home, financial contract) may be encoded into the blockchain and transacted, validated, or preserved in a much more efficient manner than at present including ideas, health data, financial assets, automobiles, and government documents."[15]

This innovation is important since it provides a way to insure the validity of a transaction in a network. Block chains can insure that data transmitted by sensor networks is valid because it meets the requirements encoded in the block chain.

10.9   The Impact of Recent Internet Innovations

Internet innovations have already had a dramatic impact on enterprise IT spending. They are expected to have an even greater impact in the future. RackSpace reports that in 2015, "68 percent of enterprises run less than a fifth of their application portfolio in the cloud; 55 percent of enterprises report that a significant portion of their existing application portfolio is not in cloud, but is built with cloud-friendly architectures."[16] This seems like slow progress. Nevertheless, Intel has estimated that by 2020, 65–85% of applications, nearly all of them, will be run on cloud infrastructure.[17]

Nomura Research[18] has confirmed these rapid changes. By 2014, many firms shifted IT consumption from traditional internal infrastructure—not based on cloud computing or software-defined infrastructure—to cloud-based infrastructure. Furthermore, by 2018, Nomura finds that spending on non-cloud-based internal infrastructure will drop to 40% of total enterprise IT consumption.[19]

Why is this happening? First, software-defined ecosystems use software, some of which is developed with Open Source and APIs, to create much less expensive and far more capable—scalable as well as interoperable—infrastructure. While these changes don't appear in US productivity data, they represent tremendous changes in the cost and ability of business to use ever more sophisticated computing, storage, and networking.

Thus, the innovations identified here are having a real effect on the economy. When firms have used this new software-defined infrastructure that permits them to implement far more mathematically sophisticated data analysis, they are saving billions of dollars by optimizing supply chains[20] and other service operations. In some cases, as at Boeing, they are expanding the scale of production[21] far beyond what was possible with traditional infrastructure.

Thus, the new infrastructure innovations described here are already having significant impacts on profits and return on assets. In the future, driverless cars and wearable devices will depend on the software-defined infrastructure described here and open vast new opportunities to expand how our economy uses computing, networking, and storage.

10.9.1   As Enterprises Adopt Cloud Computing, the Firm Changes to a "Programmable Enterprise"[22]

One illustration of how the innovations cited above are changing the economy is the shift in the way the firm operates. I call this new change a shift to the "programmable enterprise." I have chosen this phrase because the innovations in infrastructure changes mean that a firm's competitiveness and dynamism are now defined more by how it adopts the innovations enumerated in the previous sections.

We argue here that changes in technology based on the virtualization of computing and networking will transform the business world into one where software and services play a dominant role. This will change the jobs people perform and mean that most jobs will require more technical skills. These changes will occur very rapidly and traditional training mechanisms very likely will be unable to adjust to them.

The "programmable enterprise" differs from traditional firms. It is shaped by three main trends: cloud computing; handling big data; and managing sensor ecosystems that are part of the Internet of Things. New types of jobs will result from these trends. Some illustrations are: continuous service delivery jobs where employees become skilled generalists and part of DevOps teams, data analytics teams, and jobs related to sensor ecosystems.

Programmable firms have new, open, extensible, and interoperable computing and communications architectures. These businesses expect to survive based upon their ability to create new services and applications based on this new architecture. As illustrated by the MetLife case, the analysis of big data will be at the heart of "programmable enterprises."

The programmable firm represents not only talented and very skilled people, but also substantial numbers of managerial, support, and marketing employees. It is highly "people-centered" because its central purpose is connecting people and responding to their needs.

We recently estimated based on spending for cloud services that spending on cloud services (cloud computing, data analysis, and the Internet of Things) will contribute $1.7 trillion in new spending, add 3 trillion to GDP, and create about 8 million jobs for the US economy by 2025. This would add about 1.5% per year to GDP and about 0.5% per year to employment growth.[23]

10.9.2   The "Programmable Enterprise"—The Motor Promoting Dramatic Job Change

Three trends are shaping the evolution of firms:

1. The adoption of cloud computing and a shift of applications to a cloud environment.
2. Developing the ability to deal with big data because of the need to run real-time analytics and to link operational tools and processes to applications.[24] In part, this responds to much greater use of mobile applications.

3. Moving into sensor ecosystems and Internet of Things. This dominates retailing, firms with a web presence, such as eBay and Twitter, and firms that maintain and control equipment or services, including driverless cars. This trend will affect health care, but not until many operational and data privacy issues are resolved.

As a result, "programmable firms" are *distinguished by the new, open, extensible, and interoperable computing and communications architecture they have deployed and continue to expand.*

Rather than being "creative," these businesses expect to survive based upon their ability to create new services and applications for customers and suppliers. The innovative architecture—possible only with the help of refinements to virtualization and cloud computing over the last 3–5 years, turns functions and processes that had once been impossible and time-consuming to create into a key characteristic of "programmable firms." It provides them with the ability to operate new ecosystems, especially a series of functions driven by huge amounts of data analysis. These functions provide firms with the ability to develop software platforms that support innovative services of substantial value. They also offer paths to improve the value offered by such services in the future.

To illustrate this with a case, MetLife is one of the firms that has moved rapidly to establish itself as a provider of services, going beyond being a firm that merely sells insurance. It focused much of its transformation on its ability to gather large amounts of data on policyholders, not only characteristics of individuals holding policies, but also data on the places they live, the health care in their community, climate, and other dimensions.

With tools that come from the latest generation of big data analysis, MetLife created a wall for all its policyholders and their recent transactions. Using the wall, a MetLife employee can respond to a policyholder's call and see an entire window of recent interactions with the company, based upon sophisticated data management tools and links. This provides timely and comprehensive service without the usual delays.

But this is not just an easier way to respond to inquiries. It indicates that new technologies, such as Hadoop together with tools, such as MongoDB[25] are providing new ways to employ data analytics to obtain insights into many firms' big data lakes.

In addition, because of its command over big data, MetLife can do what other firms, such as Netflix already do.[26] It can see what groups of users like to do when they call for policy information. It can offer them different bundles of services or policies and see what works and what does not. It can evaluate what is going amiss when policyholders or customers terminate their policies and try to correct the flaws. It can also analyze the behavior of specific populations of policyholders or groups within the population.

This indicates that the analysis of big data and creation of new services will be at the heart of "programmable enterprises." It also indicates that what is new about these companies is:

1. A *unique architecture* in their enterprise computing and networking, based broadly upon open systems and standards, with easy "end-to-end" connectivity.[27] This architecture facilitates resource sharing and agile creation of new services.
2. Their ability to analyze and manage enormous amounts of data.
3. The ability to create new services based upon the innovative architecture and data analytics.
4. The development of *software and services platforms* within the organization and outside of it to distribute and manage existing, emerging, and future services.

Building on these innovative capabilities, the "programmable firm" represents not only talented and very skilled people, but also substantial numbers of managerial, support, and marketing employees. These employees are probably going to account for 30–60%[28] of a "programmable firm." The engineering and research, operations, IT and IS staff will account for about 40–70% of employees, with the higher share being true for startups.

In addition, through secondary impacts, the employment in these firms might provide jobs in related areas that also have substantial multiplier impacts of their own. Prof. Enrico Moretti[29] has estimated that Silicon Valley firms create a disproportionate number of legal and engineering jobs as well as many personal service jobs.

Thus, the jobs created by new "programmable firms" are likely to include more than just IT jobs. In their secondary or multiplier impacts, these firms are also more likely to support the creation of a wide range of jobs largely outside of information technology.

The new architecture for business does several things that will affect jobs and the pattern of employment:

1. It creates new jobs, such as DevOps positions that combine software design, software testing, and software deployment. This is an example of a "Hybrid" job, but it also indicates that the new jobs in the emerging economy will be far more skilled than previously. Rather than providing tasks, new jobs will demand skills.

2. It provides a way for services to create new value in the economy. The new economy is not likely to be a guild economy where the central focus is "liberating value."[30] Rather, it is innovating to create new services on top of the new computing and networking architecture, largely by exploiting new software platforms that can provide new functions for the economy.

3. Software and services become the defining features of the new economy. This means that productivity gains are based upon improvements in the efficiency in the use of resources and processes that can be controlled and managed by software.

Companies can be overwhelmed by the number of transactions they must manage and the constant demand for new applications and services due to the glut of mobile communications and a rapid rise in data flows. To respond, some firms have developed rapid action teams to make their firms more productive and responsive.

This permits such firms not only to derive important advantages from the new software-defined infrastructure that supports accelerating product and service development, but also to create work teams or groups that meld skills that had formerly been isolated to perform specific tasks. So, these firms are terminating jobs, such as software developers or quality assurance employees. The new positions they are filling in teams or squads have superseded the old occupations. The new occupational designation for these employees is not clear, nor are the tasks clearly defined. Team and squad members may be called platform team engineers but this is not true of all cases where such positions exist.

These changes in the workforce are part of an experiment in agility that has begun over the past 5 years. As the use of virtualized infrastructure with software-defined networking, cloud computing, and containers becomes more widespread, it is providing firms with a chance not only to speed the creation of new products and services, but also to emphasize creativity and teamwork above task work.

The new types of jobs that will arise from these trends are:

1. Programming jobs and support for "continuous service delivery" and the creation of microservices. These positions usually merge old technology competencies, so a "platform engineer" may need to know about DevOps (software programming), networks, and storage. In some cases, positions that formerly required only business knowledge, like a data center manager may need to be more technical and know how to work "hands-on" with Hadoop.

2. Data analytics teams that must perform rapid analyses and set up data visualization and reports on trends and new directions. This requires higher level competency with new database tools, such as Hadoop and Apache. It also requires lower level jobs to assemble and manage the data and translate the analysis into understandable discussions or reports.

3. Jobs from sensor ecosystems will expand due to several requirements. First, the building and deployment of ecosystem infrastructure, particularly computing, storage, and networking at the edge of large enterprise or public networks. These ecosystems will most likely evolve through several stages, so the building will not end in the first phase of building but require several stages of rebuilding to add more capabilities. Second, data analysts and managers to support the retailing, web, and maintenance firms that will be big early users. Major areas of expansion will be data analytics (extending big data analytics mentioned above), computer security, and software developers.

10.10 Conclusions

This chapter identifies Internet innovations made possible by the emergence of software-defined infrastructure. Most of these are based upon the fact that software is driving innovation in the way we use infrastructure and how we develop new services and applications.

Without understanding the central role of software in innovations, it would be easy to conclude that the Internet has not contributed to recent innovations in the US and global economies and will not contribute to future growth. Given the wide range of innovations identified here, that would be an unfair oversimplification.

While it is almost impossible to find the impact of these innovations in current data, they are creating a new generation of infrastructure and services that is likely to have as dramatic an impact on the economy as the early ascent of the Internet.

We also argue that changes in technology based on virtualization will dramatically change the types of jobs people perform. These changes will occur very rapidly and traditional training mechanisms will be unable to adjust. This will open new categories of jobs for people that lack formal degrees. Employers like Microsoft[31] are already moving people to the types of jobs that will be needed in this new economy and providing them with new skills to function effectively.

Notes

1. We use the phrase software-defined ecosystems as a synonym for the "New IP" and the Internet of Things. Some groups believe that the innovation described here is a precursor for much broader changes in the future. The 2nd International Workshop on Software-Defined Ecosystems (BigSystem 2015) notes: "With the emerging technology breakthrough in computing, networking, and storage, the boundary of systems is undergoing fundamental change and is expected to logically disappear. It is the time to rethink system design and management without boundaries toward software-defined ecosystems, i.e., the BigSystem. The basic principles of software-defined mechanisms and policies have witnessed great success in clouds and networking. We are expecting broader, deeper, and greater evolution of these concepts and technologies, and a confluence toward holistic *software-defined ecosystems*." [emphasis added] Association for Computing Machinery, Portland, Oregon, June 15–19, 2015, http://bigsystem.ece.ufl.edu/2015/.

2. Saunders, Steve (2014), "Introducing 'The New IP,'" *Light Reading*, September 11, http://www.lightreading.com/nfv/nfv-elements/introducing-the-new-ip-/a/d-id/710779.

3. Gordon, Robert (2016), *The Rise and Fall of American Growth* (Princeton and Oxford: Princeton University Press).

4. Andreesen, Marc (2011), "Why Software Is Eating the World," *Wall Street Journal*, August 20, http://www.wsj.com/articles/SB10001424053111190348 09045765122509156294460.

5. "Open-Source Software," *Wikipedia*, https://en.wikipedia.org/wiki/Open-source_software.

6. Hardesty, Linda (2016), "AT&T Wants 50% of Its Software to Be Open Source," *SDxCentral*, January 5, https://www.sdxcentral.com/articles/news/att-wants-50-of-its-software-to-be-open-source/2016/01/.

7. "OpenStack," *Wikipedia*, https://en.wikipedia.org/wiki/OpenStack.

8. "Open API," *Wikipedia*, https://en.wikipedia.org/wiki/Open_API.

9. Ibid.

10. Boyd, Mark (2015), "The Year Ahead: APIs as Economic Game Changers," *The NewStack*, December 28, http://thenewstack.io/year-ahead-api-game-changer/.

11. "Continuous Service Delivery in the Enterprise," *Puppet Labs*, http://www.techrepublic.com/resource-library/webcasts/continuous-service-delivery-in-the-enterprise/. Adrian Cockcroft (2015), "Cloud Trends, DevOps, and Microservices," July, http://www.slideshare.net/Indicee/cloud-trends-2015-pdf?qid=3b0ac472-a6d8-4816-97f4-b0e785015cc7&v=&b=&from_search=3.

12. "DevOps is a new term emerging from the collision of two major related trends. The first was also called "agile system administration" or "agile operations"; it sprang from applying newer Agile and Lean approaches to

operations work. The second is a much-expanded understanding of the value of collaboration between development and operations staff throughout all stages of the development lifecycle when creating and operating a service, and how important operations has become in our increasingly service-oriented world." "What Is DevOps?" http://theagileadmin.com/what-is-devops/.

13. Puppet Labs, "2015 State of DevOps Report," https://puppetlabs.com/2015-devops-report.
14. Swan, Melanie (2015), "Bitcoin and Blockchain Explained: Not just Cryptocurrencies, Economics, and Markets; Applications in Art, Health, and Literacy," April 8, http://www.slideshare.net/lablogga/bitcoin-and-block-chain-technology-explained-not-just-cryptocurrencies-economics-and-mar-kets-applications-in-art-health-and-literacy.
15. Swan, Melanie, "Bitcoin and Blockchain."
16. RightScale (2015), "RightScale Releases 2015 State of the Cloud Report," February 18, http://www.rightscale.com/press-releases/rightscale-releases-2015-state-of-the-cloud-report.
17. Waxman, Jason (2015), "Data Center Day" Intel, August 27, http://www.intc.com/events.cfm.
18. Norton, Steven (2016), "CIOs Reduce Budget Forecasts for 2016: Nomura," *Wall Street Journal*, March 22, http://blogs.wsj.com/cio/2016/03/22/cios-reduce-budget-forecasts-for-2016-nomura/.
19. Columbus, Louis (2016), "CIOs Are Prioritizing Big Data Analytics, Cloud Computing and Security in 2016/2017 Budget Cycles," *Forbes Tech*, March 23, http://www.forbes.com/sites/louiscolumbus/2016/03/23/cios-are-prioritizing-big-data-analytics-cloud-computing-and-securi-ty-in-20162017-budget-cycles/#73a6a5ff54f6.
20. Cohen, Robert (2016), Case study of a major automaker prepared for the OECD initiative on digital data innovation. Unpublished.
21. Cohen, Robert (2016), Case study of Boeing prepared for the OECD initiative on digital data innovation. Unpublished.
22. This section is adopted from an early version of my essay, "The "First Software Age": The Programmable Enterprise and Creating New Types of Jobs," in David Nordfors, Vint Cerf, and Max Senges, eds., *Disrupting Unemployment* (Kansas City: Ewing Marion Kauffman Foundation, 2016), 65–74.
23. Cohen, Robert, "Enterprise Spending on Cloud Services Will Expand US GDP, Jobs and Tech Spending. A New Forecast Predicts the US Economy will gain nearly 3 Trillion Dollars in GDP and 8 Million new Jobs from 2015 to 2025." Economic Strategy Institute, http://www.econstrat.org/research/the-new-ip-and-the-internet-of-things/638-cloud-services-till-ex-pand-us-gdp-jobs-and-tech-spending.

24. Anderson, Chris (2008), "The End of Theory Will the Data Deluge Makes the Scientific Method Obsolete?" *Edge: The Third Culture*, June 30, https://www.edge.org/3rd_culture/anderson08/anderson08_index.html. Anderson argues that scientific methods based upon "grand theories" are not likely to survive the emergence of big data. He says of the petabyte world that "It calls for an entirely different approach, one that requires us to lose the tether of data as something that can be visualized in its totality. It forces us to view data mathematically first and establish a context for it later." I would argue that we are likely to find new and innovative methods to analyze and visualize big data and I am not sure I would agree with Anderson's conclusion. I think with new mathematical and visualization techniques, we may continue to be able to visualize data in its entirety.

25. Harris, Derrick, "The promise of better data has MetLife investing $300M in new tech," *GigaOm Research*, https://gigaom.com/2013/05/07/with-300m-earmarked-for-tech-innovation-metlife-wants-to-remake-insurance/.

26. Adrian Cockcroft of Battery Ventures, formerly at Netflix, personal discussion with the author at the Open Network Users Group (ONUG), Columbia University, May 2015.

27. ONUG is a group that is championing this approach. Many of its members, including major banks and financial firms, insurance companies, retailers, and logistics firms have led the move to deploy these systems.

28. I have used Jigsaw.com profiles accessed June 2, 2015 and Codingvc.com, "Analyzing Angel List Job Postings, Part 1: Basic Stats," September 8, 2014, http://codingvc.com/analyzing-angellist-job-postings-part-1-basic-stats. See Cohen, Robert B. (2015), "A Software-Defined Economy: Innovation, the Internet and Future Growth: Describing Enterprise-based Innovations Driving Change," Columbia University CITI Conference on Internet and Employment, June 5.

29. Moretti, Enrico (2013), *The New Geography of Jobs* (Boston: Mariner Books, Houghton Mifflin Harcourt) and Brokaw, Leslie (2012), "The Multiplier Effect of Innovation Jobs," *MIT Sloan Management Review*, June 6, http://sloanreview.mit.edu/article/the-multiplier-effect-of-innovation-jobs/.

30. This essay offers a different perspective from the one by Geoffrey Moore, "Developing Middle Class Jobs in a Digital Economy," Conference presentation, Columbia University, Columbia Institute for Tele-Information, New York, NY, "The Internet's Effect on Employment," June 5, 2015. This essay proposes that the economy is moving to a new stage where computing and communications architectures alter the economy. Central to this change is the view that new value creation will no longer be dependent on creating value-adding niches along existing value chains.

31. Staten, James (2015), "It's Not the Technology, It's you," OpenStack Silicon Valley Conference, Mountain View, CA, August 26.

Bibliography

Anderson, C. 2008. "The End of Theory Will the Data Deluge Makes the Scientific Method Obsolete?" *Edge: The Third Culture*. Accessed September 30, 2017. https://www.edge.org/3rd_culture/anderson08/anderson08_index.html.

Andreessen, M. 2017. "Why Software Is Eating the World." *Wall Street Journal*, August 20, 2011. Accessed September 30, 2017. http://www.wsj.com/articles/SB10001424053111903480904576512250915629460.

Bigsystem.ece.ufl.edu. 2017. "BigSystem2015|International Workshop on Software-Defined Ecosystems." Accessed September 26, 2017. http://bigsystem.ece.ufl.edu/2015/.

Boyd, M. 2017. "The Year Ahead: APIs as Economic Game Changers." *The NewStack*, December 28, 2015. Accessed September 30, 2017. http://thenewstack.io/year-ahead-api-game-changer/.

Brokaw, L. 2017. "The Multiplier Effect of Innovation Jobs." *MIT Sloan Management Review*, June 6, 2012. Accessed September 30, 2017. http://sloanreview.mit.edu/article/the-multiplier-effect-of-innovation-jobs/.

Codingvc.com. 2014. "Analyzing Angel List Job Postings, Part 1: Basic Stats." Accessed September 30, 2017. http://codingvc.com/analyzing-angellist-job-postings-part-1-basic-stats.

Cohen, R. 2015. "A Software-Defined Economy: Innovation, the Internet and Future Growth: Describing Enterprise-based Innovations Driving Change." Columbia University CITI Conference on Internet and Employment, June 5, 2015.

Cohen, R. 2016a. "Case Study of a Major Automaker Prepared for the OECD Initiative on Digital Data Innovation."

Cohen, R. 2016b. "Case Study of Boeing Prepared for the OECD Initiative on Digital Data Innovation."

Cohen, R. 2016c. "Enterprise Spending on Cloud Services Will Expand US GDP, Jobs and Tech Spending. A New Forecast Predicts the US Economy Will Gain Nearly 3 Trillion Dollars in GDP and 8 Million New Jobs from 2015 to 2025." Economic Strategy Institute. Accessed September 30, 2017. http://www.econstrat.org/research/the-new-ip-and-the-internet-of-things/638-cloud-services-till-expand-us-gdp-jobs-and-tech-spending.

Cohen, R. 2017. "The 'First Software Age': The Programmable Enterprise and Creating New Types of Jobs." In *Disrupting Unemployment*, edited by D. Nordfors, V. Cerf, and M. Senges, 65–74. Kansas City: Ewing Marion Kauffman Foundation.

En.Wikipedia.org. 2017a. "Open API." Accessed September 30, 2017. https://en.wikipedia.org/wiki/Open_API.

En.wikipedia.org. 2017b. "Open-Source Software." Accessed September 30, 2017. https://en.wikipedia.org/wiki/Open-source_software.

En.wikipedia.org. 2017c. "OpenStack." Accessed September 30, 2017. https://en.wikipedia.org/wiki/OpenStack.

Gordon, R. 2016. *The Rise and Fall of American Growth*. Princeton and Oxford: Princeton University Press.

Hardesty, L. 2017. "AT&T Wants 50% of Its Software to Be Open Source." *SDxCentral*, January 5, 2016. Accessed September 30, 2017. https://www.sdxcentral.com/articles/news/att-wants-50-of-its-software-to-be-open-source/2016/01/.

Harris, D. 2013. "The Promise of Better Data has MetLife Investing $300M in New Tech." *GigaOm Research*, May 7. Accessed September 30, 2017. https://gigaom.com/2013/05/07/with-300m-earmarked-for-tech-innovation-metlife-wants-to-remake-insurance/.

Moore, G. 2017. "Developing Middle Class Jobs in a Digital Economy." Conference Presentation, Columbia University, Columbia Institute for Tele-Information, New York, NY, The Internet's Effect on Employment, June 5, 2015.

Moretti, E. 2013. *The New Geography of Jobs*. Boston: Mariner Books and Houghton Mifflin Harcourt.

Norton, S. 2016. "CIOs Reduce Budget Forecasts for 2016: Nomura." *Wall Street Journal*, March 22. Accessed September 30, 2017. http://blogs.wsj.com/cio/2016/03/22/cios-reduce-budget-forecasts-for-2016-nomura/.

Puppet Labs. 2017. "Continuous Service Delivery in the Enterprise." *Tech Republic*. Accessed September 30, 2017. http://www.techrepublic.com/resource-library/webcasts/continuous-service-delivery-in-the-enterprise/.

Puppet Labs. 2017. "2015 State of DevOps Report." Accessed September 30, 2017. https://puppetlabs.com/2015-devops-report.

Right Scale. 2015. "RightScale Releases 2015 State of the Cloud Report." February 18. Accessed September 30, 2017. http://www.rightscale.com/press-releases/rightscale-releases-2015-state-of-the-cloud-report.

Saunders, S. 2017. "Introducing 'The New IP'." *Light Reading*, September 11, 2014. Accessed September 30, 2017. http://www.lightreading.com/nfv/nfv-elements/introducing-the-new-ip-/a/d-id/710779.

Slideshare.net. 2017. "The Future of Cloud Innovation, Featuring Adrian Cockcroft." Accessed September 30, 2017. http://www.slideshare.net/Indicee/cloud-trends-2015-pdf?qid=3b0ac472-a6d8-4816-97f4-b0e785015c-c7&v=&b=&from_search=3.

Staten, J. 2015. "It's Not the Technology, It's You." OpenStack Silicon Valley 2015 Conference, Mountain View, CA, August 26, 2015. https://www.youtube.com/watch?v=-OJLtSxk6uo.

Swan, M. 2015. "Bitcoin and Blockchain Technology Explained: Not Just Cryptocurrencies, Economics, and Markets; Applications in Art, Health, and Literacy," April 8. Slideshare.net. Accessed September 30, 2017. http://www.slideshare.net/lablogga/bitcoin-and-blockchain-technology-explained-not-just-cryptocurrencies-economics-and-markets-applications-in-art-health-and-literacy.

The Agile Admin. 2017. "What Is DevOps?" Accessed September 30, 2017. http://theagileadmin.com/what-is-devops/.

Waxman, J. 2015. "'Data Center Day' Intel Investor Relations—Events & Presentations—Events Calendar." http://www.intc.com/event.

Part III

Polices to Facilitate Structural and Social Adjustments without Slowing Innovation

11

ICT Innovation, Productivity, and Labor Market Adjustment Policy

Robert D. Atkinson

There is increased interest in the issue of how to facilitate labor market adjustments from technological innovation. Some of this interest appears to be a result of efforts to try to respond to the recent increase in populism from both the right and the left, fueled, as some believe, by labor market insecurities. Some are due to the weak global labor market performance in the wake of the Great Recession, where tens of millions of jobs were destroyed, and job creation has been tepid. And some is due to the belief that technological change, particularly in the information and communications technology sector (ICT) is fueling (or about to fuel) rapid productivity growth and accompanying job loss. Regardless of the reasons for interest, improving worker adjustment policies in the United States is long overdue. The alternatives—doing relatively little—risks not only increasing opposition to ICT-driven technological change but reducing the efficiency of the labor market.

Before making some suggestions to improve adjustment policy it's worth first examining the relationship between innovation and jobs. While productivity growth is the main driver of increases in living standards, in the wake of the Great Recession a growing chorus of voices asserts that economies can no longer afford productivity because it kills jobs. The new narrative is that productivity driven by increasingly powerful IT-enabled "machines" is the

R. D. Atkinson (✉)
The Information Technology and Innovation Foundation,
Washington, DC, USA

© The Author(s) 2018
L. Pupillo et al. (eds.), *Digitized Labor*, https://doi.org/10.1007/978-3-319-78420-5_11

cause of slow job growth, and in the future accelerating technological change will only make things worse. Many policymakers now believe that they can't afford to support policies that boost productivity because productivity gains come at the expense of needed job growth. If productivity advances come with employment retreat, then policymakers would be well within their rights to be concerned about supporting policies to advance productivity. But fortunately, they need not worry, for there is no tradeoff.

Yet the large and growing chorus of "tech kills jobs" voices persists. Lawrence Summers recently said that he no longer believed automation would always create new jobs. "This isn't some hypothetical future possibility," he said, "This is something that's emerging before us right now."[1] Financial pundit Nouriel Roubini asks forebodingly, "Rise of the Machines: Downfall of the Economy?" Joseph Stiglitz states, "It doesn't have political appeal to say the reason we have a problem [job losses] is we're so successful in technology."[2] Paul Krugman writes: "A much darker picture of the effects of technology on labor is emerging. In this picture, highly educated workers are as likely as less educated workers to find themselves displaced."[3] Moshe Vardi, a professor at Rice University, predicts that with the development of artificial intelligence that global unemployment will reach 50%.[4] Mike Rettig of the Brookings Institution asks with mirth, "Will the last human worker please turn out the lights?"[5] In *The New Yorker*, Gary Marcus writes, "as machines continue to get smarter, cheaper, and more effective, our options dwindle. So, don't bother polishing up that resume, rather here's a link to the unemployment office."[6] Robert Reich argues that robots will "take away good jobs that are already dwindling. They will in short supplant the middle class."[7] Perhaps no one has done more to advance the idea that productivity kills jobs than MIT professors Erik Brynjolfsson and Andrew McAfee. In their popular book, *The Race Against the Machine: How the Digital Revolution Is Accelerating Innovation, Driving Productivity, and Irreversibly Transforming Employment and the Economy*, they write that workers are, "losing the race against the machine, a fact reflected in today's employment statistics."

To start with, all these statements are odd, because if technology-led productivity growth really has been the culprit behind America's anemic job growth since 2009, one would expect that America's productivity growth rate would be higher than normal. In fact, US productivity growth since the end of the Great Recession has been at historic lows—about half the rate before the Great Recession. What the pundits are attributing to anemic productivity growth actually has its roots in the painful and slow recovery from the greatest financial crisis since the Great Depression.

Indeed, historically, there has actually been a negative relationship between productivity growth and unemployment rates. In other words, higher productivity meant lower unemployment. This correlation is shown in the 2011 McKinsey Global Institute report, "Growth and Renewal in the United States: Retooling America's Economic Engine."[8] MGI looked at annual employment and productivity change from 1929 to 2009 and found that increases in productivity are correlated with increases in subsequent employment growth, and that the majority of years since 1929 feature concurrent employment and productivity gains. In looking at 71 10-year slices, only 1% had declining employment and increasing productivity. The rest showed increasing productivity and employment. In looking at 76 five-year periods, just 8% had declining employment and increasing productivity.

In the 1960s, US productivity grew 3.1% per year while unemployment averaged 4.9%. However, during the 1980s, productivity grew just 1.5% while unemployment rates averaged 7.3%. And in the 2000–2007 period, productivity was growing at a healthy 2.7% per year while the unemployment rate was under 5%. But from 2008 to 2015, productivity growth was only 1.2% yet the unemployment rate averaged over 7.5%. Moreover, recently there has been a modestly positive correlation between productivity growth and the labor force participation rate in the 34 Organization for Economic Co-operation and Development (OECD) nations from 2009 to 2015. In other words in nations with stronger productivity, more workers, not less, entered the labor force.

Today's pessimistic views that productivity kills jobs suffer not only from a lack of historical perspective, but also from a fundamental flaw in logic. That flaw is not that people who lose their jobs will get jobs making the new machines. No rational organization spends money to increase productivity unless the savings are greater than the costs. If there are the same number of jobs in the company making the machines as there are lost in the companies using the machines, then costs could not have fallen.

So, it's not that jobs will be created in the new "robot" firms, it's that they will be created across the economy from the new demand that higher productivity enables. To see how, we need to look at second-order effects, something techno-pessimists do not do. If jobs in one firm or industry are reduced or eliminated through higher productivity, then by definition production costs go down. These savings are not put under the proverbial mattress, they are recycled into the economy, in most cases though lower prices or higher wages. This money is then spent, which creates jobs in whatever industries supply the goods and services that people spend their increased savings or earnings on. As a side note, the same logic is true for profits as well. Even if all the savings went to profits, these are distributed to share-holders who in turn spend at least some of this money, creating demand that is met by new jobs. Even if the shareholders don't spend all of it, the savings reduce interest rates which leads to new capitalized spending (e.g., car loans and mortgages) and investment, which in turn creates jobs in the firms producing this additional output. Moreover, because of competitive pressures in industries, firms don't have unlimited pricing power. If they did, then firms could just raise prices now. Competitive markets force firms to pass savings along in the form of lower prices (or higher wages).

Some will argue that people won't spend the money from lower prices or higher wages, and therefore jobs won't be created. But most Americans would have little problem finding ways to spend their added income if their take-home pay increased from a doubling or even tripling of productivity. In fact, the first thing most would likely do is break out their shopping lists. To see where the new jobs from higher productivity would likely be created, we only have to look at how those in the top-income quintile spend their money versus those in the middle. According to the Bureau of Labor Statistics, top-income households spend a larger share of their income on things like education, personal services, hotels and other lodging, entertainment, insurance, air travel, new cars and trucks, furniture, and major appliances. So, if US productivity doubles, people would spend more than double on these kinds of goods and services, and employment would grow in these industries. Even if productivity were miraculously to increase by a factor of five or even ten, then the vast majority of US households would likely have no problem spending all their added income (either as personal consumption or through higher taxes for public goods, such as a cleaner environment, better cities, or more infrastructure). This is even more true in developing nations where median per-capita income is just $6000. Productivity in these nations could increase by a factor of 50 and still come nowhere near exhausting people's desires for goods and services.

As a recent study by Deloitte notes, technological innovation crates jobs in four different ways.[9] First, in some sectors where demand is responsive to price changes, automation reduces prices but also spurs more demands leading to at least compensating job creation. For example, as TV prices have fallen and quality increased, people have bought many more TVs. Second, jobs are created making the automation equipment. Workers are employed in factories making robots. Third, in some industries technology serves as a complement to workers, making output more valuable, leading to increased demand. For example, as doctors have gained better technology, the demand for health care has increased. Finally, as discussed above, reduced prices from automation increases consumers purchasing power which creates jobs at the industries they spend their new additional income on.

Not only is the notion that productivity kills jobs rebutted by history and logic, virtually all academic studies on the topic have found that productivity increases do not decrease the number of people working or raise the unemployment rate. If anything, the opposite is true. Trehan found that, "The empirical evidence shows that a positive technology shock leads to a reduction in the unemployment rate that persists for several years."[10] The OECD finds that, "Historically, the income generating effects of new technologies have proved more powerful than the labor-displacing effects: technological progress has been accompanied not only by higher output and productivity, but also by higher overall employment."[11] In its 2004 *World Employment Report*, the International Labor Organization found strong support for simultaneous growth in productivity and employment in the medium term.[12] In a paper for the International Labour Organization's *2004 World Employment Report*, Van Ark, Frankema, and Duteweerd found strong support for simultaneous growth in per-capita income, productivity, and employment in the medium term.[13] A study by Industry Canada's Jianmin Tang found that for 24 OECD nations, "at the aggregate level there is no evidence of a negative relationship between employment growth and labor productivity growth... .This finding was robust for rich or poor countries, small or large, and over the pre- or post-1995 period."[14] The United National Industrial Development Organization finds that in fact, "productivity is the key to employment growth."[15] It goes on to note:

The link between productivity and the creation of jobs is strong but somewhat complex. In a static formulation, employment and productivity are in an inverse relationship: A given quantity of work to be done will require fewer and fewer jobs as productivity increases. In dynamics, though, the relationship is altogether different. Real wages divided by labor productivity is what defines the share of the wage bill in value added. Thanks to this relationship, the share of the wage bill can be reduced without affecting the income of the workers. The larger capital residual stimulates investment and, finally, jobs.[16]

To be sure, this is not to say that in economic recessions productivity might not be accompanied by consumer demand from lower prices and job growth from increased demand, since by definition in these periods, demand is below supply. But the evidence and logic suggest that once demand returns (e.g., when the recession ends) productivity once again leads to compensating job growth. Nor is this to suggest that if productivity is higher than average in some industries—particularly industries with low elasticity of demand, where lower prices don't lead to accordingly higher sales—that it cannot lead to fewer jobs in those particular industries. But this is very different than the aggregate, economy-wide effects many doomsayers are forecasting.

In summary, even in the face of history, logic and overwhelming scholarly evidence, the "tech kills jobs" true believers remain unconvinced. Even if they acknowledge that productivity hasn't yet killed jobs, for them the future will be different. This is a seductive argument, of course, because there is no way to prove or disprove it.

The doomsayers tell a story about technological change accelerating so much that soon there will be "nowhere left to run": After the super-intelligent robots take our jobs, there will be no new jobs left to create. The narrative is as follows: As automation reduced agricultural jobs, people moved to manufacturing jobs. After manufacturing jobs were automated, they moved to service-sector jobs. But as robots automate these jobs, too, there will be no new sectors to move people into. This argument is not new. Economist Wasily Leontif warned in 1983 that:

> We are beginning a gradual process whereby over the next 30–40 years many people will be displaced, creating massive problems of unemployment and dislocation. In the last century, there was an analogous problem with horses. They became unnecessary with the advent of tractors, automobiles, and trucks. … So, what happened to horses will happen to people, unless the government can redistribute the fruits of the new technology.[17]

In 2006, Ray Kurzweil argued in *The Singularity Is Near* that because of Moore's Law, IT will remain on a path of rapidly declining prices and rapidly increasing processing power, leading to developments we can only barely imagine, such as smart robots and bio-IT interfaces.[18] Kurzweil claimed, "gains in productivity are actually approaching the steep part of the exponential curve."[19] (In fact, productivity growth rates fell by half after he wrote this.) A year later, Stuart Elliott, in a paper for the National Research Council, extrapolates Moore's Law and argues that in 23 years computers are likely to displace 60% of all jobs.[20] Five years later Brian Arthur wrote, "when farm jobs disappeared, we still had manufacturing jobs, and when these disappeared we migrated to service jobs. With this digital transformation, this last repository of jobs is shrinking—fewer of us in the future."[21] And most recently McAfee and Brynjolfsson wrote that we are "reaching the second half of the chessboard," where exponential gains in computing power lead to drastic changes after an initial gestation period.[22]

Some even go so far as to claim that artificial intelligence will lead to "superintelligence," where intelligent machines do all jobs and more, which will spell the end of jobs, and maybe even the end of the human race if the smart machines decide it is in their best interest to kill us.[23] For these pessimists, computers and robots will eclipse the full range of human ability—not only in routine manual or cognitive tasks, but also in more complex actions or decision-making. The logic is as follows: In order for there to be labor demand, there must be things that humans can do better or more cheaply than machines, but machines are becoming more useful than (a large majority of) workers in almost every conceivable way. The gloomy conclusion is we will all be living in George Jetson land (from the US TV show from the 1960s, *The Jetsons*), but unlike George, we won't be working at Spacely Sprockets, we will be at home on the dole, with only Mr. Spacely employed, because he is the one who owns the robots.

But techno-utopians make three crucial mistakes. First, as discussed below, they wrongly assume that current technological trends will continue or even accelerate. Second, they overstate the extent to which digital innovation is transforming occupations. For some of the them, virtually all jobs will be disrupted by smart machines. One of the most widely cited studies on this matter, from Osborne and Frey, found that 47% of US jobs *could* be eliminated by technology over the next twenty years.[24] But they appear to overstate this number by including occupations that have little chance of automation, like fashion models. Osborne and Frey also rank industries by the risk that their workers would be automated. They find that in accommodation and food services, "as many as 87 percent of workers are at risk of automation, while only 10 percent of workers in information are at risk."[25] While this is a speculation about the future, one would expect that there would be some positive correlation between recent productivity growth and risk of automation. In other words, industries they expect to be most at risk of being automated (by definition, through productivity growth) should have enjoyed higher productivity growth in the last few years, since many of the technologies Osborne and Frey expect to drive automation are already here, albeit not at the same levels of deployment. But in fact, there was a negative correlation between the risk of automation in an industry and the industry productivity growth of 0.26.

Moreover, even Osborne and Frey admit that "could be eliminated" is not the same as "will be eliminated." A more likely estimate is that only about 20% of US jobs are likely to be easily automated over the next decade or two, with about 50% being difficult to automate, and the remaining 30% extremely difficult to automate.[26] One reason for this difference is that, for many occupations, automation doesn't affect the occupation so much as it affects the tasks performed in an occupation. For example, the McKinsey Global Institute concludes that "Very few occupations will be automated in their entirety in the near or medium term. Rather, certain activities are more likely to be automated, requiring entire business processes to be transformed, and jobs performed by people to be redefined."[27] In other words, technology will lead much more to job redefinitions and opportunities to add more value, not to outright job destruction. If 20% of an administrative assistant's time is spent on tasks that can be automated, that doesn't mean we lose 20% of administrative assistants—it means they can spend that time doing more meaningful things instead of routine tasks such as weekly scheduling.

But even if Osborne and Frey are right and 47% of jobs are eliminated by technology over the next 20 years, this would be equivalent to an annual labor productivity rate of 4% a year, barely higher than the productivity rate of the US economy in the 1960s, when unemployment was at very low levels and job creation was high. Similarly, a Citibank report on the future of work ominously predicted that new developments in computer "algorithms could displace around 140 million knowledge workers globally."[28] This indeed might sound ominous until one realizes that this accounts for just 4.6% of global employment and any process is likely to take at least a decade or two to work its way through the labor market.

The techno-utopians third mistake is that this "nowhere left to run" argument is absurd on its face because global productivity could increase by a factor of 50 without people running out of things to buy. Just look at what people with higher incomes spend their money on: nicer vacations, larger homes, more restaurant meals, more entertainment like concerts and plays, etc. Moreover, if we ever get that rich, there would be a natural evolution toward working fewer hours. In sum, the worries of machines overtaking humans are as old as machines themselves.

That said, even if productivity is not reducing the number of jobs, isn't it making the labor market more insecure as more workers lose their jobs? This clearly seems to be what most workers think. In 1987, a solid majority of US workers (59%) said they felt their jobs were secure; by 2014, less than half felt that way (47%).[29] Yet while people feel less secure now than in the past, employment data tell a different story. Data from the U.S. Bureau of Labor Statistics clearly disprove the idea that average American workers are trapped in a perpetual state of job insecurity, regardless of how much they may happen to earn. In fact, Americans today are less likely to lose their jobs than they were in the 1990s. Looking at the broadest measures of total job loss—defined as jobs eliminated when an establishment closes down or downsizes—the US economy has seen fewer jobs lost as a share of total employment, with similar trends at the individual industry level. Because each establishment is a single physical location that either produces goods or provides services, a single business may have one or more establishments. US workers in 1995 had around a 7.3% chance that their jobs would be eliminated in any given quarter. Two decades later, that figure was down to 5.7%.[30]

The same trend of greater job security holds across industries. Of 10 major sectors, all saw a lower rate of job loss in 2015 than in 1995 defined as the share of jobs lost in that industry through contractions or closings. However, job security differs across industries. For example, in 1995, roughly 15% of jobs per quarter were lost in the construction industry, while the education and health services sectors eliminated about 5% of jobs. Nonetheless, the general trend is reduced losses. Consider that if the share of job losses remained unchanged from 1995 levels, the manufacturing sector would have incurred about two million additional worker displacements in 2015. In fact, while neither manufacturing output nor employment has yet to recover to 2007 levels, compared with all other economic sectors, the risk of losing one's job is the lowest of all major sectors.

So, the evidence is clear that higher productivity from ICT does not lead to net job loss, nor is worker insecurity up. Still, for many workers and advocates the right level of labor market disruption is zero. No one should ever lose their job. Of course, the problem with this is that by definition innovation is about making some industries and occupations redundant, what Schumpeter famously referred to as creative destruction. We didn't need very many buggy whip makers after the car. Over the last decade, we have needed a lot fewer travel agents after online travel booking. And in the future, we can be sure that many occupations, including some currently thought to be relatively immune from disruption from ICT, will in fact be disrupted.

Indeed, radically new models of service delivery could expand and emerge, significantly disrupting many occupational labor markets. Imagine the number of university professors falling from 1.7 million down to perhaps 500,000 as most students take high-quality massively open online courses (MOOC). Imagine advanced software tools providing many of the services now provided by personal investment advisors and business benefit advisors. Imagine autonomous vehicles reducing the number of long-haul truck drivers and taxi drivers. Exactly how this process will take place, at what velocity, and in what industries and occupations is, of course, harder to predict. Indeed, as noted innovation economist Joseph Schumpeter wrote, "Technological possibilities are an uncharted sea." But what we can chart is that the waters will not be calm.

However, efforts to reduce creative destruction would not only reduce innovation and the rate of productivity growth, it would do little to help workers. In the United States while workers were more likely to lose their job in the 1990s from firm downsizing or closures than since 2009, their labor market fortunes were better in the 1990s, with reduced average length of unemployment and higher wages for their next job.[31]

Moreover, there is strong evidence that stricter labor market regulations designed to protect workers from disruption have a large negative impact on ICT investment and the benefits firms can obtain from it. Van Reenen et al. find that labor market regulations reduce productivity gains from ICT by approximately 45%.[32] The authors attribute one third of this effect to how labor market regulations can slow down the entry and exit of firms: stricter regulations can protect and preserve less productive, less technologically advanced firms. Labor market regulations also reduce the flexibility of managers, preventing them from reorganizing production in more efficient ways. Why buy IT to reorganize production and cut costs when regulations make it difficult to reduce the workforce? Antonelli similarly finds that rigid labor markets make firms less likely to adopt ICTs.[33]

So, if the answer is not to resist ICT-driven innovation, it likewise can't be simply embracing flexible labor markets with workers on their own to adjust to disruption. One promising direction will be to do a better job of ensuring that students have more ICT skills. Despite the views of many, demand for computer science is not consigned just to IT professions. As Ed Lazowska, the Bill and Melinda Gates Chair in Computer Science and Engineering at the University of Washington, states: "Every field is becoming an information field, and if you can program at a level beyond an intro course, it's a huge value to you."[34] Demand for computer knowledge is ubiquitous, and transforms traditional sectors across the economy. Many occupations, suggests IT expert David Moschella, now require "double-deep" skills, with training and expertise in technology and computing in addition to the skills traditionally demanded by these occupations.[35] In today's technology-fueled economy, most industries rely on computer skills. Two-thirds of computer jobs are in non-technology industries, such as healthcare, banking, or manufacturing.[36] Organizations are increasingly technology-driven and technology dependent. For marketers, managers, bankers, designers, accountants, and others, coding experience and advanced understandings of computing technology are increasingly valuable. Professionals are learning technology and analytical skills and IT specialists are applying their focused skills onto a wide range of practical business applications.[37] Similarly, workers in middle-skilled manufacturing jobs have a need for computer and technology skills. Workers with advanced computer knowledge who can use their experience to address and solve a host of problems and challenges are poised in succeed in a wide variety of fields. This means doing more to support computer science education, particularly in high schools and colleges.

Unfortunately, only around a quarter of high schools offer computer science, and often this course lack rigor or focus on computer use or just coding instead of delving into computer science principles. Only 18% of schools accredited to offer Advanced Placement exams offer the computer science AP exam. And only 22% of students who take the AP exam in computer science are female, the largest gender disparity of any AP exam.[38] Moreover, access to computer science is also limited at universities, where institutions limit enrollment through restrictions, higher admission standards, or introductory "weed-out" courses designed to keep students out of the major. In many cases, universities have few incentives to incur the cost of expanding computer science programs in response to student demand. These artificial constraints disproportionally impact women and minorities, diminishing attempts to promote inclusivity.

To address these challenges policymakers should reform curricula for existing technology classes to focus on core concepts of computer science in primary and secondary schools and provide resources to train and recruit high-quality computer science teachers. All states should allow computer science to count as either a math or science requirement, and more STEM-intensive public high schools that give students in-depth exposure to computer science should be established to allow students with the aptitude and interest in computer science to more deeply explore the subject. And universities should be incentivized to expand their offerings in computer science and prioritize retaining students interested in majoring, minoring, or taking courses in computer science.

But we need to go beyond just computer science education. When worker skills are more developed worker adjustment from dislocation becomes easier.[39] And one key way workers get needed skill is through on the job training. However, corporate investment in workforce training has declined significantly in the past two decades, and that is a big problem for American productivity and international competitiveness. As the *Economic Report of the President* finds, the proportion of workers that received employer-sponsored training dropped 42% between 1996 and 2008.[40] And despite the rhetoric that workers are the main priority for companies, corporate spending on training as a share of gross domestic product (GDP) declined from more than half a percent in 2000 down to one-third of a percent in 2013.[41] These cuts have made it harder for workers to find new employment after they are laid off and have made it more difficult for US firms to boost productivity and global competitiveness.

Corporations have cut their investment in workforce training for a number of reasons. Declines in employee tenure in the 1980s and 1990s meant that more and more firms sought to simply hire workers with the requisite skills instead of paying to train them. After all, why invest in human capital development when that asset will likely walk out the door to a competitor firm before the investment pays off? The increasing focus on short-term profits has also driven corporations to invest less in the future than they did previously.

In short, this is a classic case of market failure. Firms invest less in training than is optimal from a societal and economic perspective and it negatively impacts economic growth and innovation. It's the same reason firms invest less in research and development than is societally optimal. To fix the latter problem, Congress created the research and experimentation (R&E) tax credit in 1983 to incentivize companies to spend more on research and development (R&D). We need to follow the same model here. Congress should turn the R&E credit into a knowledge tax credit by allowing qualified expenditures on both R&D and workforce training to be taken as a credit. Under the current alternative simplified R&D credit, firms can claim 14% of all expenditures above 50% of base period expenditures. To ensure that companies use this credit to focus on the skills of the majority of their workers, and not just managers, firms taking advantage of the credit would need to abide by rules similar to those for pension program distribution, which limit focus on highly compensated employees.

Federal policy needs to do a better job at ensuring that education is better linked to occupational needs, particularly for middle-skill jobs. One highly successful program designed to build technician skills is NSF's Advanced Technological Education (ATE) program, which supports community colleges working in partnership with industry, economic development agencies, workforce investment boards, and secondary and other higher education institutions. ATE projects and centers are educating technicians in a range of fields, including nanotechnologies and microtechnologies, rapid prototyping, biomanufacturing, logistics, and alternative fuel automobiles. Notwithstanding this, ATE funding is quite small, at around $50 million per year. Congress should expand funding for the ATE program to at least $100 million per year.

In addition, federal policy should do more to help establish wider use of skills credentialing systems. The National Skill Standards Act of 1994 created a National Skill Standards Board (NSSB) responsible for supporting voluntary partnerships in each economic sector that would establish industry-defined national standards leading to industry-recognized, nationally portable certifications. The vision was that each industry define and validate national standards for the skills it was seeking and credential individuals against those skills. One key reason for doing this was so that companies would have a better way to assess the skills of prospective and current workers and so that workers would have a better way to identify and gain the skills they need to be successful. But while some industries stepped up to the plate to organize such a system through the Manufacturing Skill Standards Council (MSSC), the federal government failed to provide matching funding to establish this standards-based system. Moreover, in the 2000s, the national approach was abandoned in favor of a regional approach (embodied in programs such as the Department of Labor's Employment and Training Administration's WIRED—Workforce Innovation for Regional Economic Development—initiative) which contributed to an uncoordinated proliferation of certifications at the regional and state levels. What's really needed is a national approach, so that employers can more readily find workers with the right skills for advanced manufacturing and workers can be confident their skills will be recognized similarly by employers across the entire country. Therefore, Congress and the Administration should work to increase credentialing by expanding the use of standards-based, nationally portable, industry-recognized certifications specifically designed for specific sectors.

The rise of Internet job matching platforms also can play a role in helping adjustment. These platforms often provide needed work and income for workers in transition between jobs. Well-known gig platforms include Uber and Lyft (ride sharing), UpCounsel (legal experts), Instacart (shopping and delivery), and TaskRabbit (odd jobs). All use a combination of Internet and mobile technology to match workers with consumers. One challenge however is that existing labor law makes provides an all or nothing system with regard to the platform-worker relationship. If the platform engages in activities like training, withholding taxes and other services, they increase the chance that courts will find the existence of an employer-employee relationship, which brings with it a host of other obligations. As a result, most gig platforms err on the side of not providing these services to their workers.

One solution would be for Congress to create a special exemption from many of the labor laws specifically for gig platforms. Platforms are unique enough that legislation could define them fairly precisely, making it clear whom the law covers and whom it does not. Despite their rapid growth, they are also a small enough part of the workforce that treating them differently would not upend the broader labor markets. An exemption, even if it lasted only 5 or 10 years, would give Congress a chance to experiment with the application of labor laws to a new century. The temporary nature could motivate firms to provide more services to their workers in order to persuade Congress to extend and broaden it. We could see whether companies are willing to create a more supportive and involved relationship with their workers in order to reduce turnover, improve quality, and enhance their public reputations. We could also see whether these attempts actually benefitted workers and raised their incomes or job satisfaction.

Finally, policy needs to do more to help workers who lose their jobs. One path to not take is what many continental European nations take, paying workers who lose their jobs relatively high payments for relatively long periods of time. For example, in France and Germany unemployed workers, even ones fired for misconduct, can receive benefits for two years at relatively high levels of wage replacement.[42] Not only do these generous policies hurt job creation—by paying workers not to work they reduce consumer demand from the rest of the workforce who must pay higher taxes to support the generous unemployment insurance payments—but they contribute to an atrophy of skills and an increased duration of unemployment.[43] In other words, the longer a worker is unemployed, the lower their chances of exiting unemployment and reduces their wages when they finally become reemployed.[44]

At the same time, limited benefits and leaving dislocated workers on their own is not an answer either. In the United States, the level of unemployment insurance benefits largely depends on the state in which the worker lives, and the variation in benefits is quite significant, with workers in some states like Mississippi and Arkansas receiving approximately one third the benefits of workers in states like New Jersey and Washington. The challenge therefore is to increase benefits without reducing the incentive for workers to get back in the workforce.[45] One solution is for Congress to increase the Federal Unemployment Tax Act (FUTA) that employers must pay, so that the unemployment insurance tax and benefit floor across the nation increases, so that the UI benefits in third of the states providing the lowest benefits increases. At the same time, the Department of Labor should design incentives so that state unemployment insurance programs provide benefits that decline with the length of unemployment. In other words, the initial amount of benefit would be higher than it is now, but would decline gradually by perhaps 5% for every two weeks being unemployed. This could be done in a benefits neutral way so that the average worker would still receive the same amount of benefits but would now have a stronger incentive to find work.

Finally, policy needs to do more than enable laid off workers to gain the skills they may need to get back into the workforce. Unfortunately, in many states unemployed workers must be available for work to get benefits and being enrolled in a certified training program can disqualify them for benefits. In other words, just when a worker is available to gain new or upgraded skills (when they are unemployed) policy is often preventing that from happening. Congress could mandate that states change these restrictive policies.

Related to this, Congress should do more to help workers who lose their jobs from technological change. Since the 1960s the United States has had Trade Adjustment Assistance Act (TAA) program. The program was designed in part of substantive reasons to help workers hurt by trade, but also to reduce opposition to trade by helping those hurt by trade. As President Kennedy stated in 1962 at when he signed TAA legislation, "When considerations of national policy make it desirable to avoid higher tariffs, those injured by that competition should not be required to bear the full brunt of the impact. Rather, the burden of economic adjustment should be borne in part by the Federal Government." Today it is time to adapt and expand TAA into a comprehensive Trade and Technology Adjustment Assistance Act (TTAA), to help all displaced workers, no matter the cause of their displacement—and to help workers adapt to changes brought by gains in productivity and automation.[46]

In conclusion, the major risk to the global economy over the next decade is not too much disruption, but too little. In other words, the risk is that productivity will grow too slowly. As such it is critical that labor market policies, including adjustment policies, support, not hinder ICT-led creative disruption. One way to do that is do a better job at workforce training and labor market adjustment policies.

Notes

1. Summers, Lawrence H. (2014), "What Does the Future Hold for Our Economy," (Panel Discussion at Aspen Ideas Festival 2014; Uploaded to YouTube on July 1), http://www.aspenideas.org/session/what-does-future-hold-our-economy, Accessed March 7, 2016.
2. Condon, Bernard, and Paul Wiseman (2013), "Recession, Tech Kill Middle-Class Jobs," *Associated Press*, January 23, http://bigstory.ap.org/article/ap-im-pact-recession-tech-kill-middle-class-jobs, Accessed March 7, 2016.

3. Krugman, Paul (2013), "Sympathy for the Luddites," *The New York Times*, June 13, http://www.nytimes.com/2013/06/14/opinion/krugman-sympathy-for-the-luddites.html?_r=0.

4. Yuhas, Alan (2016), "Would You Bet Against Sex Robots? AI 'Could Leave Half of World Unemployed,'" *The Guardian*, February 13, http://www.theguardian.com/technology/2016/feb/13/artificial-intelligence-ai-unemployment-jobs-moshe-vardi.

5. Rettig, Mike (2015), "Will the Last Human Worker Please Turn Out the Lights?" *The Hill*, September 21, http://thehill.com/blogs/pundits-blog/labor/254337-will-the-last-human-worker-please-turn-out-the-lights.

6. Marcus, Gary (2012), "Will a Robot Take Your Job?" *The New Yorker*, December 29, http://www.newyorker.com/news/news-desk/will-a-robot-take-your-job.

7. Reich, Robert (2015), *Saving Capitalism: For the Many, Not the Few* (New York: Alfred Knopf).

8. Manyika, James, David Hunt, Scott Nyquist, Jaana Remes, Vikram Malhotra, Lenny Mendonca, Byron Auguste, and Samantha Test (2011), "Growth and Renewal in the United States: Retooling America's Economic Engine," (McKinsey Global Institute: February), http://www.mckinsey.com/global-themes/americas/growth-and-renewal-in-the-us, Accessed March 8, 2016.

9. Stewart, Ian, Debapratim De, and Alex Cole (2015), "Technology and People: The Great Job-Creating Machine," (Working Paper, Deloitte), http://www2.deloitte.com/content/dam/Deloitte/uk/Documents/finance/deloitte-uk-technology-and-people.pdf, Accessed March 10, 2016.

10. Trehan, Bharat (2003), "Productivity Shocks and the Unemployment Rate," *Federal Reserve Bank of San Francisco Economic Review*, http://www.frbsf.org/economic-research/files/article2.pdf, Accessed March 7, 2016.

11. Organisation for Economic Co-Operation and Development (OECD) (1998), *Technology, Productivity and Job Creation: Best Policy Practices* (Paris: OECD) 9, http://www.oecd.org/dataoecd/39/28/2759012.pdf, Accessed March 7, 2016.

12. International Labour Organization (2005), *World Employment Report 2004–05: Employment, Productivity, and Poverty Reduction* (Geneva: ILO).

13. van Ark, Bart, Ewout Frankema, and Hedwig Duteweerd (2004), "Productivity and Employment Growth: An Empirical Review of Long and Medium Run Evidence" (Working Paper, Groningen Growth and Development Centre, May).

14. Tang, Jianmin (2015), "Employment and Productivity: Exploring the Trade-off," *Industry Canada*, http://www.csls.ca/ipm/28/tang.pdf, Accessed March 3, 2016.

15. Isaksson, Anders, Thiam Hee Ng, and Ghislain Robyn (2005), *Productivity in Developing Countries: Trends and Policies* (Vienna: UNIDO), 139.

16. Ibid., 138.

17. Leontief, Wasily, and Faye Duchin (1984), "The Impacts of Automation on Employment, 1963–2000," *New York Institute for Economic Analysis*, April, http://eric.ed.gov/?id=ED241743, Accessed March 7, 2016.
18. Kurzweil, Ray (2006), *The Singularity Is Near: When Humans Transcend Biology* (New York: Penguin Books).
19. Ibid.
20. Elliott, Stuart W. (2007), "Projecting the Impact of Computers on Work in 2030," (Presentation Paper, National Research Council).
21. Arthur, W. Brain (2011), "The Second Economy," *McKinsey Quarterly*, October, http://www.mckinsey.com/business-functions/strategy-and-corporate-finance/our-insights/the-secondeconomy, Accessed March 7, 2016.
22. McAfee, Andrew, and Eric Brynjolfsson (2011), *Race Against the Machine: How the Digital Revolution Is Accelerating Innovation, Driving Productivity, and Irreversibly Transforming Employment and the Economy* (Digital Frontier Press, October 17).
23. Bostrom, Nick (2015), *Super Intelligence: Paths, Dangers, Strategies* (Audible Studios on Brilliance Audio, May 5).
24. Frey, Carl Benedikt, and Michael A. Osbourne (2013), "The Future of Employment: How Susceptible Are Jobs to Computerisation?" (Oxford Martin School, University of Oxford, Oxford, September 17), http://www.oxfordmartin.ox.ac.uk/downloads/academic/The_Future_of_Employment.pdf, Accessed March 8, 2016.
25. Citi GPS: Global Perspectives and Solutions, "Technology at Work v2.0: The Future Is Not What It Used to Be," (Oxford Martin School, University of Oxford, Oxford, January 2016), 60, http://www.oxfordmartin.ox.ac.uk/downloads/reports/Citi_GPS_Technology_Work_2.pdf, Accessed March 8, 2016.
26. Miller, Ben (2013), "Automation Not So Automatic," *The Innovation Files*, September 20, http://www.innovationfiles.org/automation-not-so-automatic/, Accessed March 7, 2016.
27. Chu, Michael, James Manyika, and Mehdi Miremadi (2015), "Four Fundamentals of Workplace Automation," *McKinsey Quarterly*, November.
28. Citi GPS, "Technology at Work v2.0," 10.
29. Adams. Susan (2014), "Most Americans Are Unhappy at Work," *Forbes*, June 20, https://www.forbes.com/sites/susanadams/2014/06/20/most-americans-are-unhappy-at-work/#2fcef98d341a.
30. Atkinson, Robert D., and John Wu (2016), "The U.S. Labor Market Is Far More Stable Than People Think" (Information Technology and Innovation Foundation, June), https://itif.org/publications/2016/06/20/us-labor-market-far-more-stable-people-think.
31. Ibid.
32. Van Reenen, John et al. (2010), "The Economic Impact of ICT, SMART" (Centre for Economic Performance), http://www.ukn.inet, Accessed June 23, 2016.

33. Antonelli, Cristiano, and Francesco Quatraro (2013), "Localized Technological Change and Efficiency Wages Across European Regional Labour Markets," *Regional Studies* 47 (10): 1686–1700.

34. Soper, Tyler (2014), "Analysis: The Exploding Demand for Computer Science Education, and Why America Needs to Keep up," *Geekwire*, June 6, http://www.geekwire.com/2014/analysis-examining-computerscience-education-explosion/.

35. Moschella, David (2013), "The Emerging Double-Deep Economy," *Leading Edge Forum*, September 17, https://leadingedgeforum.com/publication/the-emerging-double-deep-economy-2318/.

36. "Make Computer Science in K-12 Count!" Code.org, Computing in the Core, Accessed January 11, 2016, https://code.org/files/convince_your_school_or_state.pdf; Steve Taylor, "Obama's 'TechHire' Follow Lead of Mission' EDC's 'Code the Town,'" *Rio Grande Guardian*, March 22, 2015; http://riograndeguardian.com/obama-takes-missions-code-the-town-nationwide/.

37. Moschella, David (2013), "Double-Deep and How Its Forces Are Reshaping Today's Economy," *Leading Edge Forum*, April 4; https://leadingedge-forum.com/publication/double-deep-and-how-its-forces-arereshaping-todays-economy-2294/.

38. Nager, Adams, and Robert D. Atkinson (2016), "The Case for Improving U.S. Computer Science Education" (Information Technology and Innovation Foundation, May), https://itif.org/publications/2016/05/31/case-improving-us-computer-science-education.

39. Zimmer, Timothy E. (2016), "The Importance of Education for the Unemployed," *Indiana Business Review*, Spring, http://www.ibrc.indiana.edu/ibr/2016/spring/article2.html.

40. Furman, Jason, Maurice Obstfeld, and Betsey Stevenson (2015), "The 2015 Economic Report of the President," *News Release*, February 19, https://www.whitehouse.gov/blog/2015/02/19/2015-economic-report-president.

41. Atkinson, Robert D. (2016), "Restoring Investment in America's Economy" (Information Technology and Innovation Foundation, June), https://itif.org/publications/2016/06/13/restoring-investment-americas-economy.

42. Centre Des Liaisons Européennes Et Internationales De Sécurité Sociale (2016), The French Social Security System, Accessed June 16, http://www.cleiss.fr/docs/regimes/regime_france/an_5.html; European Commission, Employment, Social Affairs & Inclusion, http://ec.europa.eu/social/main.jsp?catId=1111&intPageId=2565&langId=en, Accessed June 16, 2016.

43. Conefrey, Thomas, Yvonne McCarthy, and Martina Sherman (2013), "Re-Employment Probabilities for Unemployed Workers in Ireland" (Working Paper, Central Bank of Ireland, Economic Letter Series, Vol. 2013, No. 6), https://www.centralbank.ie/publications/Documents/Reemployment%20Probabilities%20051113.pdf.

44. Schmieder, Johannes F., Till von Wachter, and Stefan Bender (2013), "The Causal Effect of Unemployment Duration on Wages: Evidence from Unemployment Insurance Extensions," (Working Paper, NBER, CEPR, IAB and IZA, November), http://www.econ.ucla.edu/tvwachter/papers/SchmiederVonwachterBender.pdf.

45. 2015–2016 Maximum Weekly Unemployment Benefits by State, Saving2Invest, Accessed June 16, 2016. http://www.savingtoinvest.com/maximum-weekly-unemployment-benefits-by-state/.

46. Atkinson, Robert D. (2015), "How Certain Are You That Robots Won't Create as Many Jobs as They Displace," *The Christian Science Monitor*, December 3, http://www.csmonitor.com/Technology/Breakthroughs-Voices/2015/1203/How-certain-are-you-that-robots-won-t-create-as-many-jobs-as-they-displace; United States Department of Labor, Employment and Training Administration, "What Is Trade Adjustment Assistance?" last updated June 22, 2012, https://www.doleta.gov/tradeact/factsheet.cfm.

References

2015–2016 Maximum Weekly Unemployment Benefits by State, Saving2Invest. Last Accessed October 1, 2017. http://www.savingtoinvest.com/maximum-weekly-unemployment-benefits-by-state/.

Adams, Susan. 2014. "Most Americans Are Unhappy at Work." *Forbes*, June 20. Last Accessed October 1, 2017. https://www.forbes.com/sites/susanadams/2014/06/20/most-americans-are-unhappy-at-work/#2fcef98d341a.

Antonelli, Cristiano, and Francesco Quatraro. 2013. "Localized Technological Change and Efficiency Wages Across European Regional Labour Markets." *Regional Studies* 47 (10): 1686–1700.

Arthur, W. Brain. 2011. "The Second Economy." *McKinsey Quarterly*, October. Last Accessed October 1, 2017. http://www.mckinsey.com/business-functions/strategy-and-corporate-finance/our-insights/the-secondeconomy.

Atkinson, Robert D. 2015. "How Certain Are You That Robots Won't Create as Many Jobs as They Displace." *The Christian Science Monitor*, December 3. Last Accessed October 1, 2017. http://www.csmonitor.com/Technology/Breakthroughs-Voices/2015/1203/How-certain-are-you-that-robots-won-t-create-as-many-jobs-as-they-displace; United States Department of Labor, Employment and Training Administration. 2012. "What Is Trade Adjustment Assistance?" Last Updated June 22. Last Accessed October 1, 2017. https://www.doleta.gov/tradeact/factsheet.cfm.

Atkinson, Robert D. 2016. "Restoring Investment in America's Economy." Information Technology and Innovation Foundation. https://itif.org/publications/2016/06/13/restoring-investment-americas-economy.

Atkinson, Robert D., and John Wu. 2016. "The U.S. Labor Market Is Far More Stable Than People Think." Information Technology and Innovation Foundation. https://itif.org/publications/2016/06/20/us-labor-market-far-more-stable-people-think.

Bostrom, Nick. 2015. *Super Intelligence: Paths, Dangers, Strategies*. Audible Studios on Brilliance Audio, May 5.

Centre Des Liaisons Européennes Et Internationales De Sécurité Sociale, The French Social Security System. Last Accessed June 6, 2016. http://www.cleiss.fr/docs/regimes/regime_france/an_5.html; European Commission, Employment, Social Affairs & Inclusion. Last Accessed June 16, 2016. http://ec.europa.eu/social/main.jsp?catId=1111&intPageId=2565&langId=en.

Chu, Michael, James Manyika, and Mehdi Miremadi. 2015. "Four Fundamentals of Workplace Automation." *McKinsey Quarterly*, November.

Citi GPS: Global Perspectives and Solutions. 2016. "Technology at Work v2.0: The Future Is Not What It Used to Be." Oxford Martin School, University of Oxford, Oxford, 60. Last Accessed October 1, 2017. http://www.oxfordmartin.ox.ac.uk/downloads/reports/Citi_GPS_Technology_Work_2.pdf.

Condon, Bernard, and Paul Wiseman. 2013. "Recession, Tech Kill Middle-Class Jobs." *Associated Press*, January 23. Last Accessed March 7, 2016. http://bigstory.ap.org/article/ap-impact-recession-tech-kill-middle-class-jobs.

Conefrey, Thomas, Yvonne McCarthy, and Martina Sherman. 2013. "Re-Employment Probabilities for Unemployed Workers in Ireland." Working Paper, Central Bank of Ireland, Economic Letter Series, 06/EL/13.

Elliott, Stuart W. 2007. "Projecting the Impact of Computers on Work in 2030." Presentation Paper, National Research Council.

Evidence from Unemployment Insurance Extensions. Working Paper, NBER, CEPR, IAB and IZA, November 2014. Last Accessed October 1, 2017. http://www.econ.ucla.edu/tvwachter/papers/SchmiederVonwachterBender.pdf.

Frey, Carl Benedikt, and Michael A. Osbourne. 2013. "The Future of Employment: How Susceptible Are Jobs to Computerisation?" Oxford Martin School, University of Oxford, Oxford, September 17. Last Accessed October 1, 2017. http://www.oxfordmartin.ox.ac.uk/downloads/academic/The_Future_of_Employment.pdf.

Furman, Jason, Maurice Obstfeld, and Betsey Stevenson. 2015. "The 2015 Economic Report of the President." *News Release*, February 19. Last Accessed March 1, 2016. https://www.whitehouse.gov/blog/2015/02/19/2015-economic-report-president.

International Labour Organization. 2005. *World Employment Report 2004–05: Employment, Productivity, and Poverty Reduction*. Geneva: ILO.

Isaksson, Anders. 2005. *Thiam Hee Ng, and Ghislain Robyn, Productivity in Developing Countries: Trends and Policies*, 139. Vienna: UNIDO.

Krugman, Paul. 2013. "Sympathy for the Luddites." *The New York Times*, June 13. Last Accessed October 1, 2017. http://www.nytimes.com/2013/06/14/opinion/krugman-sympathy-for-the-luddites.html?_r=0.

Kurzweil, Ray. 2006. *The Singularity Is Near: When Humans Transcend Biology*. New York: Penguin Books.

Leontief, Wasily, and Faye Duchin. 1984. "The Impacts of Automation on Employment, 1963–2000." New York Institute for Economic Analysis, April. Last Accessed October 1, 2017. http://eric.ed.gov/?id=ED241743.

"Make Computer Science in K-12 Count!" Code.org. Last Accessed January 11, 2016. https://code.org/files/convince_your_school_or_state.pdf; Steve Taylor. 2015. "Obama's 'TechHire' Follow Lead of Mission' EDC's 'Code the Town.'" *Rio Grande Guardian*, March 22. Last Accessed January 11, 2016. http://riograndeguardian.com/obama-takes-missions-code-the-town-nationwide/.

Manyika, James, David Hunt, Scott Nyquist, Jaana Remes, Vikram Malhotra, Lenny Mendonca, Byron Auguste, and Samantha Test. 2011. "Growth and Renewal in the United States: Retooling America's Economic Engine." McKinsey Global Institute, February. Last Accessed October 1, 2017. http://www.mckinsey.com/global-themes/americas/growth-and-renewal-in-the-us.

Marcus, Gary. 2012. "Will a Robot Take Your Job?" *The New Yorker*, December 29. Last Accessed March 1, 2017. http://www.newyorker.com/news/news-desk/will-a-robot-take-your-job.

McAfee, Andrew, and Eric Brynjolfsson. 2011. "Race Against the Machine: How the Digital Revolution Is Accelerating Innovation, Driving Productivity, and Irreversibly Transforming Employment and the Economy." Digital Frontier Press, October 17.

Miller, Ben. 2013. "Automation Not So Automatic." *The Innovation Files*, September 20. Last Accessed October 1, 2017. http://www.innovationfiles.org/automation-not-so-automatic/.

Moschella, David. 2013a. "Double-Deep and How Its Forces Are Reshaping Today's Economy." *Leading Edge Forum*, April 4. Last Accessed March 1, 2017. https://leadingedgeforum.com/publication/double-deep-and-how-its-forces-arereshaping-todays-economy-2294/.

Moschella, David. 2013b. "The Emerging Double-Deep Economy." *Leading Edge Forum*. September 17. Last Accessed March 1, 2017. https://leadingedgeforum.com/publication/the-emerging-double-deep-economy-2318/.

Nager, Adams, and Robert D. Atkinson. 2016. "The Case for Improving U.S. Computer Science Education." Information Technology and Innovation Foundation, May. https://itif.org/publications/2016/05/31/case-improving-us-computer-science-education.

Organisation for Economic Co-Operation and Development (OECD). 1998. *Technology, Productivity and Job Creation: Best Policy Practices*, 9. Paris: OECD. Last Accessed October 1, 2017. http://www.oecd.org/dataoecd/39/28/2759012.pdf.

Reich, Robert. 2015. *Saving Capitalism: For the Many, Not the Few*. New York: Alfred Knopf.

Rettig, Mike. 2015. "Will the Last Human Worker Please Turn Out the Lights?" *The Hill*, September 21. Last Accessed March 1, 2017. http://thehill.com/blogs/pundits-blog/labor/254337-will-the-last-human-worker-please-turn-out-the-lights.

Schmieder, Johannes F., Till von Wachter, and Stefan Bender. 2013. "The Causal Effect of Unemployment Duration on Wages: Evidence from Unemployment Insurance Extensions." IZA Discussion Paper No. 8700.

Soper, Tyler. 2014. "Analysis: The Exploding Demand for Computer Science Education, and Why America Needs to Keep Up." *Geekwire*. Last Accessed March 1, 2017. http://www.geekwire.com/2014/analysis-examining-computerscience-education-explosion/.

Stewart, Ian, Debapratim De, and Alex Cole, "Technology and People: The Great Job-Creating Machine." Working Paper, Deloitte.

Summers, Lawrence H. 2014. "What Does the Future Hold for Our Economy." Panel Discussion at Aspen Ideas Festival; Uploaded to YouTube on July 1, 2014. Last Accessed March 1, 2016. http://www.aspenideas.org/session/what-does-future-hold-our-economy.

Tang, Jianmin. 2015. "Employment and Productivity: Exploring the Trade-off." *Industry Canada*, vol. 28. Last Accessed October 1, 2017. http://www.csls.ca/ipm/28/tang.pdf.

Trehan, Bharat. 2003. "Productivity Shocks and the Unemployment Rate." Federal Reserve Bank of San Francisco Economic Review. Last Accessed March 7, 2016. http://www.frbsf.org/economic-research/files/article2.pdf.

Van Ark, Bart, Ewout Frankema, and Hedwig Duteweerd. 2004. "Productivity and Employment Growth: An Empirical Review of Long and Medium Run Evidence." Working Paper, Groningen Growth and Development Centre, May.

Van Reenen, John et al. 2010. "The Economic Impact of ICT, SMART." Centre for Economic Performance. Last Accessed October 1, 2017. http://www.ukn.inet.

Yuhas, Alan. 2016. "Would You Bet Against Sex Robots? AI 'Could Leave Half of World Unemployed.'" *The Guardian*, February 13. Last Accessed October 1, 2017. http://www.theguardian.com/technology/2016/feb/13/artificial-intelligence-ai-unemployment-jobs-moshe-vardi.

Zimmer, Timothy E. 2016. "The Importance of Education for the Unemployed." *Indiana Business Review*, Spring. Last Accessed October 1, 2017. http://www.ibrc.indiana.edu/ibr/2016/spring/article2.html.

12

Ensuring the Education and Skills Needed for ICT Employment and Economic Growth

Richard N. Clarke

12.1 Overview

Information and communication technologies (ICTs) are becoming an increasingly important component of modern economies. Use of these technologies has been a principal, if not the most important, driver of economic growth over the past half century. They have enabled economies to transform from a physical labor manufacturing focus into ones based on the use of intellectual knowledge to provide services.

But associated with this transformation in productive techniques and outputs has been a radical change in required labor skills. Industrial manufacturing or office work during the first three quarters of the twentieth century required mostly blue and pink collar skills: laborers or skilled tradesmen and secretaries, clerks or typists. As ICTs developed and increased their influence over both the manufacturing and service sectors, this began to change. Rather than the physical laborer being the most important person on the assembly line, it became the designer or programmer of the manufacturing robot or other automation equipment, as well as the skilled worker who maintained and repaired this

The analyses and conclusions developed in this chapter are those of the author alone, and should not be construed as representing any official position of AT&T.

R. N. Clarke (✉)
AT&T – Global Public Policy, Washington, DC, USA

© The Author(s) 2018
L. Pupillo et al. (eds.), *Digitized Labor*, https://doi.org/10.1007/978-3-319-78420-5_12

201

equipment. Offices and service production centers also were depopulated of secretaries, clerks, typists and mid to lower managers who had compiled paper reports and provided communication links between upper management and production workers. Instead, these tasks devolved upon computer and communication networks—and the persons who knew how to design, program and operate them.

Further, the skills needed to develop or use ICTs generally require education or training that differs substantially from the education and training required to be productive in the earlier twentieth century economy. Rather than trade apprenticeships, interpersonal relationships and on-the-job training, ICTs demand an educational background in science, technology, engineering and mathematics (STEM)—one which liberal-arts focused secondary and post-secondary educational institutions often struggle to provide on an efficient basis.

This chapter examines the growth and economic significance of ICTs and ICT employment in today's economy. It goes on to detail the particular skills and education required for workers to be productive participants in ICT-centric economies. It then discusses several education and training innovations that hold promise for workers to acquire efficiently these skills. Finally, it concludes by describing how one major ICT-focused corporation, AT&T, is promoting these ICT-based methods as a way to ensure that it has a productive workforce—and to provide an easy and efficient gateway for its business and residential customers to access the STEM skills necessary to make productive use of the ICT services that AT&T produces.

12.2 Influence of ICTs in Today's Economy

ICTs have a profound influence on today's economy. Not only is there a huge industry responsible for creating these technologies, but the ICT-creating industry is dwarfed by the segments of modern economies that make use of ICTs. Indeed, the influence of ICTs on the economy is much greater from their use than from their initial development and sale. But in addition to their vast influence on output markets, the unique demands of ICTs cause them to have an equally immense influence on markets for productive input factors, particularly labor. The following section will address the influence of ICTs on both output markets and labor markets.

12.2.1 What Are ICTs?

ICTs are multifaceted. Perhaps the most well-known embodiment of ICTs is in information processing equipment: computers. While in the past, computers were bulky mainframe equipment—housed in special climate-controlled rooms and tended to by a small cadre of specialist technicians, today this represents a minority of computing equipment. Today's typical computer is a desktop or laptop PC, at its largest—and goes down from there. Smartphones, digital cameras, GPS units, and even microwave ovens or clock radios, are all computers. But these are typically tended to by people whose principal profession is not computing, but who commonly have at least some knowledge or competence in the field.

Perhaps less visible than computing equipment is communications equipment. In the past, these were special purpose analog or digital switches or transmission equipment—also housed in special climate-controlled rooms or buildings and tended to by a small cadre of specialist technicians. But today, communications equipment is everywhere—not just inter-linking the large nodes of communications companies with each other and with their direct customers. Ordinary homes contain elaborate Ethernet, MoCA or Wi-Fi networks, and commercial communications use of the airwaves is everywhere—from mobile wireless to RFID sensor networks to satellite radio or television.[1]

Further, almost all present-day computing equipment also incorporates communications capabilities. The era of the standalone computer fed by punch cards and computer tapes, and outputting miles of paper printouts is over. Inputs to and outputs from today's computers now travel almost exclusively via digital communications technologies.

But ICTs are not just embodied in computing and communications equipment. Dwarfing this investment in equipment is investment in the software technologies that make this equipment operational and useful. This investment divides into two types. The first is the human investment sunk into developing the software that both operates the vitals of the ICT equipment as well as performs the applications that users intend the equipment to execute. But even more significant is the human investment in the users of this operating and applications software. While the former investment may be restricted to employees of the software companies creating these programs, the latter is diffused among all companies and individuals that actually use this equipment. Given the immenseness of this human investment, it is little wonder that operating and applications software frequently has useful lifespans many times longer than the equipment upon which it operates.

12.2.2 Influence of ICTs on Economic Growth

ICT has three major pathways of influence on the economy: via the provision of physical ICT infrastructure, via the development and provision of ICT-based applications, and via the use of ICT infrastructure and applications by businesses and residential users.

The most direct and obvious influence of ICTs on the economy is via the production of ICT infrastructure. ICT infrastructure includes items like Internet Protocol networks, electronic computing and communications equipment, the user devices connected to the network and the special-purpose building structures that may house this equipment. In 2013, US companies invested $94 billion in computers and their peripheral equipment, invested $110 billion in communications equipment, and invested $13 billion in communications structures (U.S. Department of Commerce 2015). But these communications networks and computers are of little value without operating software to bring them to life and without applications software to make them useful for the tasks that business and residence customers seek to accomplish. Investment expenditures on this software amounted to $299 billion in the US in 2013. Altogether, these ICT investments have accounted for roughly 20% of all private fixed investment in the US in recent years. The growth history of each of these categories of ICT investment is displayed in Fig. 12.1. While investment in ICT equipment and structures has been relatively stable over the past ten years, investment in software has expanded significantly.

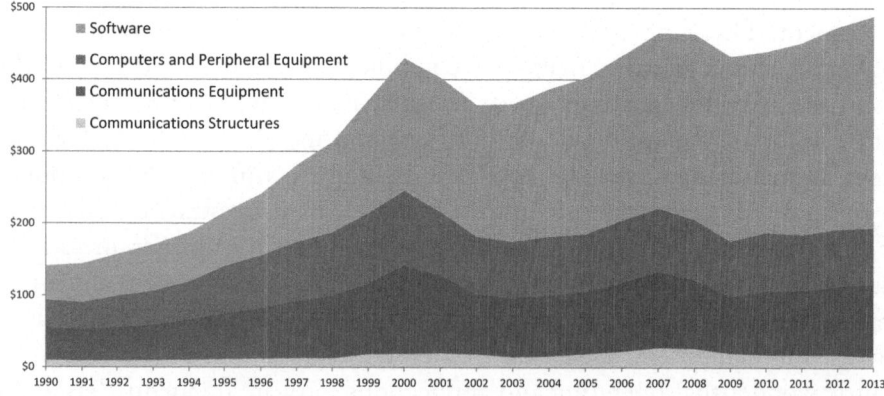

Fig. 12.1 Annual investment in ICTs in the United States: 1990–2013 (billions) (*Data source* Brogan 2014)

While these infrastructure and software investments by themselves amount to over 3% of annual US Gross Domestic Product, ICT investments represent only the tip of ICT iceberg. The most profound economic impact from ICTs derives not from the production of computers, communication networks or software, but from the use of these ICT components by businesses and residences in their productive activities. This is because the percentage of all businesses and consumers using ICTs is very much larger than the percentage of businesses creating them.

The influence of all of this ICT investment on the US national economy is nothing short of massive. Over the thirty years between 1974 and 1994, ICTs are estimated to have accounted for roughly 50% of total labor productivity growth in the US economy. While from 2004 to 2012 this share has dropped to around 40%, ICT's influence on total US growth remains very strong (Byrne et al. 2013; Miller and Atkinson 2014). Indeed, software alone is estimated to be responsible for over 15% of all US labor productivity growth between 2004 and 2012 (Shapiro 2014).

12.2.3 Influence of ICTs on Employment and Wage Growth

While it is clear that ICT production and use have had massive, positive influences on overall economic growth, their effects on labor employment and wages are more complex. This is because ICT use may displace non-ICT services and labor—which may result in some economic sectors and their workers suffering employment loss or wage reductions. Thus, the immediate effect of expanded use of ICTs on total employment and average wages may be ambiguous.

Note, first, that ICT-centric jobs exist within both ICT-producing sectors (e.g., businesses that manufacture computers or communications equipment and firms producing software or operating or installing networks) as well as in ICT-using sectors such as finance or aerospace. Note, further, that ICT-producing businesses also employ many workers engaging in non-ICT-centric jobs. As a result, ICT-related employment consists of both these groups. This is illustrated in the following graphic (Fig. 12.2).

As the above graphic also indicates, ICT occupations both inside and outside of the ICT-producing sector are growing rapidly—with a total of 7.0 million jobs in 2010 estimated to rise to 8.0 million by 2020. In addition, the growth of the ICT-producing sector has resulted in the increased employment of other occupations in the sector—rising from 3.8 million jobs in 2010 to an estimated 4.2 million in 2020 (Brogan 2012, 2013).

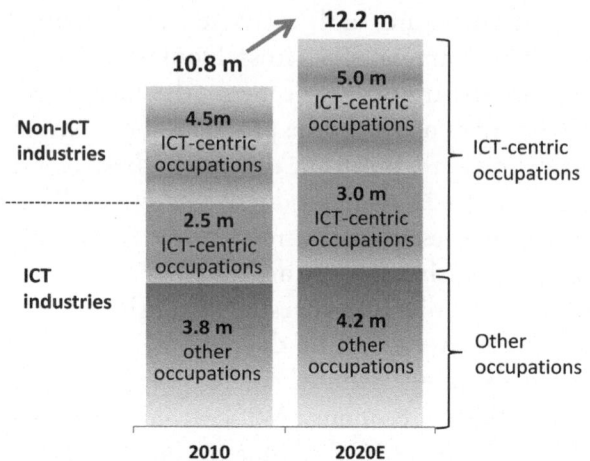

Fig. 12.2 Distribution and growth of US ICT-related occupations (*Data source* Brogan 2012)

Currently, ICT-related jobs account for over 7.5% of all employment in the United States, and this may be expected to expand because ICT-related industry sectors are among the fastest growing in the United States. Indeed, six of the eight fastest growing US industry sectors (computer manufacturing, software publishing, computer systems design, data processing and information services, securities and financial, and semiconductors) are ICT-focused. Further, seven out of the twenty US industry sectors projected to have the largest absolute amounts of output growth over the 2012–2022 period are ICT-related sectors.

As a result of this dynamic growth, not only do ICT-related jobs account for 7.5% of national employment in the United States, they pay far better than national averages. Wages for all occupations employed within ICT industries boast wage levels that are 137% of the US average wage—and 152% of the US median wage. These pay premiums are even more pronounced when one looks exclusively at ICT-centric occupations within US industries. Wages for ICT-centric occupations are 141% of the overall US average, and 176% of the US median (Brogan 2012).

Of course, this employment growth will not occur unless there are adequate numbers of workers trained and prepared to staff these ICT-related jobs—which overwhelmingly will require some form of post-secondary education for entry. Indeed, nineteen out of the thirty occupations projected to grow the fastest from 2012 to 2022 will require such post-secondary education for entry (U.S. Department of Labor 2015).

12.3 Educational Requisites for ICT-Related Employment

If workers were interchangeable between ICT-related jobs and other jobs, the impact of the ICT revolution on labor markets would be minimal. Workers would simply flow laterally from declining sectors of non-ICT employment into the faster-growing ICT sectors. But this simple movement is unlikely to be the case. ICT-related positions are much more likely to require a post-secondary educational background and STEM knowledge than other jobs in the economy. This double requirement presents a huge challenge to labor markets and to educational institutions. Further, enhanced education is not only necessary for the workers whose jobs involve ICTs. This education is also extremely valuable to enable ordinary consumers to use ICTs effectively. Thus, the fact that an enhanced education, particularly in STEM fields, makes it more likely for a worker to find a job, for the job to pay well and for the worker and his or her family to take advantage of ICTs in their home life; makes STEM education triply valuable for the economy and society.

But just as enhanced STEM education provides a triple benefit to workers and their families, it provides a double benefit to the ICT-related companies that employ them. This is because not only are such workers more productive, they are also more likely to buy and use at home the ICT-related technologies and services that their employers produce. Indeed, this virtuous cycle is akin to that attributed to Henry Ford who in 1914 doubled the wages he paid to his automotive factory workers to $5 per day. By this simple action Ford not only gained access to more productive workers—but their employment now provided them with an income adequate for them to afford to buy one of the Model T cars they were producing—and thereby increased Ford's sales (Nilsson 2014; Cwiek 2014).

12.3.1 Challenges Facing the Current Educational System

STEM education is widely provided by today's educational system, but it presents this system with many challenges. First, current high school and college degree requirements may not require the collection of many STEM credits—often as few as four classes over the course of a four-year educational program. Given that student minds tend to be far more plastic in accepting new forms of knowledge in their adolescent and young adult years than subsequently, failure to acquire substantial STEM knowledge while young and still in school is a lost opportunity that is far more difficult to recover from than to address early.

Second, STEM education as currently presented is among the most expensive for schools to convey. While mathematics may require minimal investments in school infrastructure beyond books, a lecture room and a blackboard; science, technology and engineering education commonly requires substantial physical infrastructure. Laboratory facilities need to be provided, as well as significant computing facilities. Further, while an ordinary lecture class may involve as little as three or four hours of classroom occupancy a week, laboratory classes may require much more than this per student. These laboratory and computing facilities are also not static, they may require continual renewal and reinvestment as technologies evolve.

Third, traditional STEM degrees, especially at the post-secondary level, tend to be broad programs—often covering a far wider range of topics and skills than a student may need for any particular job they eventually acquire. While acquiring these skills pre-emptively while in school may allow the student a wider selection of possible jobs when entering the labor market, it also means that a substantial fraction of the student's learning may end up being unexploited in their acquired position, and thus possibly wasted.

Finally, STEM educators are scarce and expensive to acquire. Because the United States produces so few STEM graduates relative to other countries, those that it does produce can command substantial wage premiums—as educational institutions need to bid not only against other educational institutions but also against commercial ICT-creating and using industries for this talent (National Association for Alternative Certification 2009).

12.3.2 Possible Solutions to These Educational Challenges

While all of the above factors may create a "perfect storm" threatening the increased provision of STEM-educated workers, the technology that the ICT industry creates may provide some promising ways to address these challenges.

For example, the shortage of STEM teachers may possibly be addressed by the development of Massive Online Open Courses (MOOCs). By using technology to address a class of thousands, rather than the few hundred that is likely to be the maximum capacity of a physical lecture hall, the productivity of STEM educators may be multiplied many fold. MOOCs also may hold the promise of allowing students to be taught by higher quality teachers. Educators commonly span a wide quality range, from lower quality ones who fail to communicate well and to motivate students, to higher quality ones who are superb communicators and provide students with real inspiration. ICTs that allow a higher quality teacher to direct a MOOC may increase the average teacher quality experienced by students.

ICTs also enable online learning. This, too, may result in wider and less expensive conveyance of STEM education. Learning online reduces the need for expensive schoolrooms. It also reduces the amount of transportation and time needed by students to attend classes. But perhaps most importantly, it allows time-shifting. By allowing the student flexibility over the time during which he or she chooses to "attend" class, conflicts with work schedules or childcare are reduced substantially. All of this reduces the "cost" to the student of gaining the desired education.

Innovative ICTs may also permit online learning to rival or exceed the immersiveness of an in-class education. By allowing the student to proceed at his or her own pace, and to pause the presentation to pose questions or conduct side investigations as issues arise—rather than after the lecture or physical class is over—may result in students that have better apprehended and assimilated the course's contents. Further, because of the ability of computer-generated graphics to demonstrate relationships or simulate physical experiments individually in front of each student, it is possible that this form of information presentation may be more impressionable upon students than demonstrations or experiments performed at the front of a traditional lecture hall.

Another manner in which ICTs may permit post-secondary students to acquire STEM skills at less cost is via nanodegrees (Mehrotra 2014). These programs differ from traditional university-like degrees and courses in that they are far more compressed in time and coverage. While a traditional university course runs for three or four months, a nanodegree course generally runs for only one to three months. And because nanodegree programs are far more focused in subject area than traditional degree programs, they are typically completed in six to twelve months—rather than the two to four years for traditional post-secondary degrees. Finally, because these nanodegree courses are taught via MOOC format, they are also highly economic—often costing only several hundred dollars per course—rather than the several thousand dollar price tag for traditional post-secondary STEM courses.

12.4 AT&T Educational Initiatives

AT&T is a large telecommunications and video distribution company. Every year AT&T hires roughly 25,000 new employees, nearly 75% of which will work in STEM fields (Anderson 2014). In addition improving the quality of AT&T and its business partners' workforces, better and wider STEM education will also improve the ability of consumers to use the ICT-related services that AT&T produces. To help in improving STEM education, AT&T takes a two-pronged approach. One prong is to provide support for general improvements in secondary or high school education. The other is to support specific post-secondary STEM career skills acquisition and improvement.

12.4.1 General Educational Support

Aspire is the name of AT&T's initiative to support general secondary education in the United States. This program is run through the AT&T Foundation and is committed to provide $350 million in funds over the 2008–2017 period.[2] Aspire's projects cover a broad range of initiatives, including programs to improve secondary education directly, as well as to reduce high school dropout rates and to provide students with mentoring opportunities from AT&T employees.[3]

Certain Aspire initiatives try to capitalize on the fact that you rarely need to nag an adolescent to pay more attention to his or her computer, mobile or gaming device. By funding the development of learning materials or programs that take advantage of modern mobile, broadband or video technologies and communications to convey information in an engaging manner, Aspire seeks to harness the capabilities of these ICT technologies to improve secondary education.

But in addition to improving the education that students receive when they attend school, it is vital that young people do attend school and not drop out. For this reason, Aspire also supports programs offered by organizations such as GradNation or Communities in Schools which provide out-of-school, community-based programs for at-risk students aimed at increasing their high school graduation rates.[4]

Finally, there is nothing like one-on-one or small group exposure to actual working people and their jobs to demystify what is involved with "having a job" or to relate why schooling is valuable to achieving this end. For this reason, thousands of AT&T employees donate their time to meet with secondary school students, allow them to "shadow" them in their jobs for a day, and answer questions about things like what high school classes they have found to have been the most valuable for performing their current responsibilities, what a job interview consists of, or any number of other questions whose answers may seem obvious to someone already within the world of work, but like a black box to a young student. In all, over 100,000 high school students a year participate in AT&T's job shadow program. Many of these students who go on to college may also participate in summer internship programs that AT&T sponsors.

12.4.2 Post-secondary STEM Education Initiatives

Once education reaches the post-secondary level, today's STEM education tends to become much more specialized, and generally a lot more expensive—both in terms of direct cost as well as the opportunity costs incurred by both the student and possibly the student's employer when productive labor is foregone for extended periods while pursuing a traditional college-level or graduate degree. Clearly, any educational innovations that can reduce study time and expense while providing the student with the specific STEM-skills needed to further their employment will be highly welcome.

For the above reasons, AT&T Aspire supports a nanodegree project called Udacity.[5] This project, which is a joint initiative of AT&T, Google, Facebook and other leading ICT companies, offers nanodegrees in multiple ICT-related subject areas such as web and app development, digital entrepreneurship, programming, and many others. As noted earlier, these courses are highly economic—both in terms of direct cost and students' (and possibly their employers') time. Udacity nanodegree courses only take a month or two to complete—with a full multi-course program typically being accomplished within six months to a year. Further, the cost of each course is only around $200. As a result, large numbers of leading companies like GE or CapitalOne have agreed to accept credentials offered by these Udacity nanodegrees (Schacht 2016).

For more advanced professional training, AT&T and Udacity have partnered with Georgia Institute of Technology to develop a full 36 credit-hour Master of Science in Computer Science curriculum. This program, offered via MOOC format, costs students between $7000 and $8000 in tuition to complete the full degree—only about one-fifth of what a traditionally-taught computer science Master's program would cost.[6]

Because nanodegrees are a relatively recent innovation in educational technology, large-scale data are not yet available demonstrating their degree of efficacy, but initial anecdotal reports suggest they are a success (Fenton 2015; Yegulalp 2015).[7]

12.5 Summary

There is no question but that over the last forty years, ICTs have been the most important source of economic growth and sectoral change in both US output and labor markets. Indeed, it is hard for anyone under the age of 50 to remember a time when manual processes were the exclusive way of conducting a financial transaction or making an airline reservation.

But with these changes in technology have come great changes in the educational requisites for workers. STEM knowledge and advanced education are now the key to productive employment and a middle class lifestyle. While traditional educational systems have faced difficulties in providing widespread STEM education on an efficacious basis, ICTs themselves may offer at least a partial solution. By harnessing the ability of ICTs to convey STEM education via MOOCs or nanodegrees, it may be possible to ensure that the economic revolution that ICTs have wrought in modern society will continue well through the twenty-first century.

Notes

1. Ethernet is a computer networking technology used in both local and metropolitan area networks (https://en.wikipedia.org/wiki/Ethernet). MoCA or Multimedia over Coax Alliance is a home networking technology that uses the coaxial cable already present in many houses for video and digital data distribution (http://www.mocalliance.org/about/technology.htm). Wi-Fi is the well-known wireless networking technology used for short-range communications in homes and businesses (https://en.wikipedia.org/wiki/IEEE_802.11). RFID or Radio-Frequency Identification is a wireless technology used to identify and track objects (https://en.wikipedia.org/wiki/Radio-frequency_identification).
2. https://www.att.com/Common/about_us/files/pdf/aspire/att_aspire_flyer.pdf.
3. http://www.about.att.com/content/csr/home/possibilities/at-t-aspire.html.
4. See http://gradnation.americaspromise.org/ and https://www.communitiesinschools.org/.
5. https://www.udacity.com/ and http://about.att.com/content/dam/csr/11_18assets/PCassets/Aspire_nanodegree_one-pager_11-21-14.pdf.
6. http://www.omscs.gatech.edu/.
7. See also https://www.quora.com/Are-Udacity-nanodegrees-worth-it-for-finding-a-job.

References

Anderson, Nicole. 2014. "AT&T Pledges to Support 100Kin10, National Network to Grow STEM Initiatives." November 20. http://about.att.com/content/csr/home/blog/2014/11/at_t_pledges_to_supp.html.

Brogan, Patrick. 2012. "Broadband and ICT Ecosystem Directly Supports Nearly 11 Million High-Paying U.S. Jobs." *USTelecom Research Brief*, February 28. https://www.ustelecom.org/sites/default/files/documents/022812_Employment-Research-Brief-final.pdf.

Brogan, Patrick. 2013. "Broadband and Tech Jobs Drive Economic Future." *USTelecom Blog*, December 20. http://www.ustelecom.org/blog/broadband-and-tech-jobs-drive-economic-future.

Brogan, Patrick. 2014. "Migration to Modern Networks: What Do the Latest Data Show?" *USTelecom Webinar*, August 13. http://origin-qps.onstreammedia.com/origin/InfiniteConferencing/Web%20Recordings/DatedRecordings/081314/USTelecom/081314USTelecom-Edited.mp4.

Byrne, David M., Stephen D. Oliner, and Daniel E. Sichel. 2013. "Is the Information Technology Revolution Over?", March. http://papers.ssrn.com/sol3/papers.cfm?abstract_id=2240961.

Cwiek, Sarah. 2014. "The Middle Class Took Off 100 Years Ago ... Thanks to Henry Ford?" *All Things Considered, NPR*, January 27. http://www.npr.org/2014/01/27/267145552/the-middle-class-took-off-100-years-ago-thanks-to-henry-ford.

Fenton, William. 2015. "Udacity Review & Rating." *PCMag*, May 29. https://www.pcmag.com/article2/0,2817,2484810,00.asp.

Mehrotra, Anushka. 2014. "Could Online Nanodegrees Replace 4-Year College Degrees?" *USA Today—College*, June 23. http://college.usatoday.com/2014/06/23/could-online-nanodegrees-replace-4-year-college-degrees/.

Miller, Ben, and Robert D. Atkinson. 2014. "Raising European Productivity Growth Through ICT." Information Technology and Innovation Foundation Whitepaper, June. https://itif.org/publications/2014/06/02/raising-european-productivity-growth-through-ict.

National Association for Alternative Certification. 2009. "STEMing the Teacher Shortage Tide." November. https://www.uschamberfoundation.org/sites/default/files/publication/edu/STEMing%20the%20Teacher%20Shortage%20Tide_FINAL.pdf.

Nilsson, Jeff. 2014. "Why Did Henry Ford Double His Minimum Wage?" *The Saturday Evening Post*, January 3. http://www.saturdayeveningpost.com/2014/01/03/history/post-perspective/ford-doubles-minimum-wage.html.

Schacht, Jack. 2016. "Does a Nanodegree Really Guarantee a Job?" *My College Planning Team Blog*, August 16. http://mycollegeplanningteam.com/2016/08/nanodegree-really-offer-guaranteed-job/.

Shapiro, Robert J. 2014. "The U.S. Software Industry: An Engine for Economic Growth and Employment." Software and Industry Information Association Whitepaper. https://www.siia.net/Admin/FileManagement.aspx/LinkClick.aspx?fileticket= yLPW0SrBfk4%3D&portalid=0.

U.S. Department of Commerce, Bureau of Economic Analysis. 2015. "National Income and Product Accounts." Tables 5.3.6, 5.4.6 and 5.5.6. Webpage visited December. http://www.bea.gov/iTable/iTable.cfm?ReqID=9&step=1#reqid= 9&step=1&isuri=1.

U.S. Department of Labor, Bureau of Labor Statistics. 2015. "Projections of Occupational Employment, 2014–24." December. http://www.bls.gov/career-outlook/2015/article/projections-occupation.htm.

Yegulalp, Serdar. 2015. "Even at Half Off, Is a Udacity Nanodegree Worth It?" *Infoworld Tech Watch*, July 7. https://www.infoworld.com/article/2944070/certifications/udacitys-half-off-deal-for-online-certifications-bonus-or-bust.html.

13

Smart Organizations, New Skills, and Smart Working to Manage Companies' Digital Transformation

Andrea Iapichino, Amelia De Rosa and Paola Liberace

13.1 The Transformation of Work Through Digital Transformation

The current transformation of work should be regarded as part of a bigger picture: the so-called "digital transformation" is giving birth to a complex intertwining of connections, involving both people and smart objects. Change is coming at a much faster pace than in the past: The web as we knew it quickly widens to wrap up common things, now able to provide, receive and elaborate information in order to perform autonomous tasks; sensors and memories enable everyday's objects to trace, react to and predict events. Men and machines interact in technologically dense environments,[1] thus enacting what has been called "sociomateriality".[2] Smart objects, smart homes, smart mobility, smart infrastructures and facilities, build up smart cities, as responsive and connected as their citizens and administrators. In such background, information is the new currency: it leverages shared networks and devices, coming over all exchanges and negotiations, whether individual or social, personal or professional, human- or machine driven,

A. Iapichino
University of Florence, Florence, Italy

A. De Rosa (✉)
Luiss Guido Carli University, Rome, Italy

P. Liberace
La Sapienza University, Rome, Italy

© The Author(s) 2018
L. Pupillo et al. (eds.), *Digitized Labor*, https://doi.org/10.1007/978-3-319-78420-5_13

and producing tons of data to process and analyze in order to generate added value.

Yet the transformation of work has been an outcome of technological as well as social changes:

- On the one hand, the fourth wave of the industrial revolution[3] leads to mass customization, both through advanced digitization, the availability of big, complex, real-time data, and technologies, such as cloud computing, 3D printing and additive manufacturing. This entails more reactive, flexible, autonomous production processes all across different businesses[4]: as a matter of fact, borders separating industries are fading, and manufacturing and services are about to be considered as a unique smart industry. The "platform" business model steps forward, leveraging technology in order to achieve maximum production efficiency both from customer's and from the worker's point of view. The customer becomes an empowered "prosumer", actively contributing to provide information about his/her expectations, hence enabling platform output to immediately fit them; on the other side, the "proworker", doesn't just accomplish fixed tasks, instead fulfills his/her goals through a "fuzzy" behavior, autonomous and not reproducible. Supply and demand by now meet on the platform, far from the old-fashioned "analogic" market.
- On the other hand, the financial and economic crisis experienced all across Europe has highlighted employment threats—resulting in huge job losses and insecurity—as well as new opportunities. It has assigned a major impact to work-life balance, ranging from the need for reshaping the role of work up to downshifting trends.[5] At the same time, a new consciousness has arisen about technology and its role—both empowering and invading—in daily personal and social life. Last but not least, the shift from ownership to access[6] has involved several dimensions of social life, from housing to transportation, from tourism and travel to media consumption: the so-called "sharing economy" entails the deregulation of former branches, disrupting traditional social and legal barriers.

The balance of this essay is organized in three sections. In the next section, we explain why a new, smart organization is required to cope with companies' digital transformation. The following section then focuses on the skills called by this transformation. The final section discusses the transition to a new way of working—smart working—and concludes the chapter.

13.2 Smart Organization

In this context, a new kind of "smart" organization is rapidly developing, marked by transitory layout and agile structure. Managerial control issues lose weight,[7] while output validation becomes a major issue, basically depending on the prosumer's feedback. Traditional hierarchy therefore fades, while appointments and tasks are more and more linked to specific processes and goals which entail specific skills and seniorities. Factory workload is designed and distributed following dynamic patterns, according to prosumers' needs and expectations and to proworkers' skills and availability. According to this point of view, employees' work could become totally "on demand",[8] and could potentially be divided between multiple employers; as a consequence, loyalty and devotion to the company could be replaced by different qualities, such as the ability to integrate seamless with specific, ever-changing processes.

Such a scenario is due to have a huge impact on education and training, putting skills on the foreground and challenging organizations to enhancing, improving and reconvert workers' expertise. The new organization will most likely leverage formal as well as informal learning, treasuring personal, nonconventional training experiences and peer-to-peer education. Corporate knowledge should get ready to be built, amended and shared through open platform, according to a "wiki" model; similarly, skills assessment will be based on a peer-reviewed mechanism, yet improved in order to avoid misrepresentations. A possible way to manage the assessment issue could be a decentralized network, in which data related to workers' training, evaluation and endorsement are secured with nodes, with a Blockchain-like model.

Above all, in order to succeed in the new organization, a basic set of skills is required, as in Fig. 13.1.

Reframing is essential in order to understand an ever-changing organizational environment; proworkers should be able to redefine their role on-the-go, building a scalable, flexible job profile. Empowerment goes with an entrepreneurial, autonomous and mindful behavior; it implies the ability to delegate basic, routinary tasks to machines while keeping managerial, cognitive tasks, and to take advantage of personal learnings and nonconventional training experiences in a professional viewpoint. Awareness is related to technology, acknowledging its importance in organizational life and its part in facilitating work-life balance, but also to information and data management, whether personal or professional, in order to deal with them according to civil law and corporate rules. Integration is key to manage at

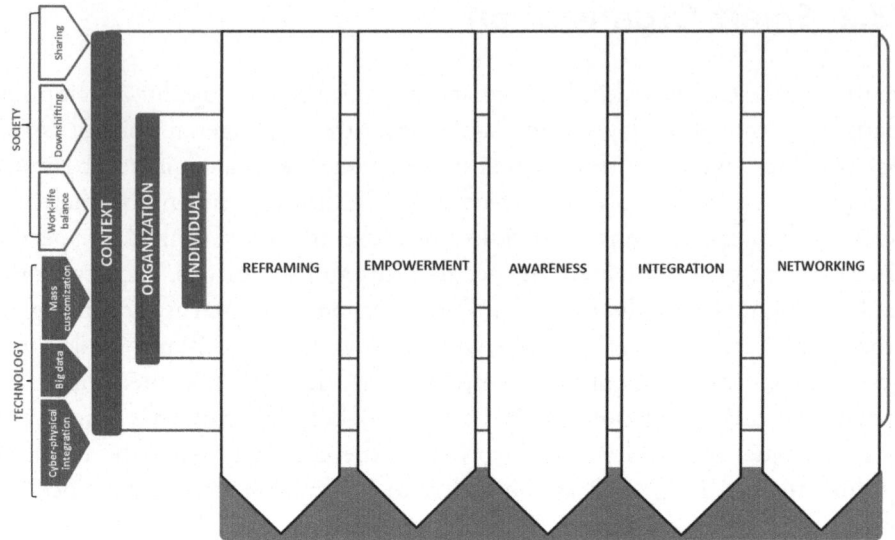

Fig. 13.1 New skills for the digitally transforming organization

once different goals, tasks and positions in the organization, entering different processes in a seamless way. Networking concerns the ability to work together in complex man-machine systems, interacting with cyber-physical networks while embracing individual, organization and context in a cross perspective.

Clarified what we mean for «digital transformation», this essay shows how TIM—as to 2016—is leveraging "knowledge management" and "smart working" to develop skills to support the digital organization.

13.3 Transition from Knowledge to New Skills: TIM Academy

We live in a "knowledge society" in which knowledge plays a central role, economically, socially and politically. This knowledge is constantly nourished through research and innovation. Therefore, universities and research centers are called to promote the enhancement and the direct application of knowledge generated through collaboration with businesses and other local stakeholders.

Here in TIM we strengthened our relationship with leading research centers and national and international universities, with whom we have activated more than 40 MOUs and agreements, in order to enhance talent, develop new capabilities and transfer innovation in the company. It's a double path which on the one hand improves our ability to innovate and on the other hand helps to develop skills.

The development and growth of internal competences are part of the strategy and the company's success, including new and innovative ways of managing our knowledge. We engaged in lessons on Wiki platforms and Open Knowledge models, implementing a continuous learning process in which all stakeholders can provide, review and amend contributions.

TIM Academy starts its journey in 2016, by introducing new models of knowledge sharing and working closely with observatories and internal and external research and innovation centers. It is both a physical and virtual place dedicated to the entire population (approx. 53,000 employees) of T.I. Group in Italy. The faculty is composed by full-time "digital-social educators" and a large community of employees which is engaged in transferring and sharing their know-how and disseminating digital skills needed to support the transformation and evolution of the company in the new technological, business and cultural environment.

"The complexity of a chip, measured for example by the number of transistors per chip, doubles every 18 months". With reference to Moore's first law, this is where companies start to face the challenge of the digital age. They must transform the "hypertrasformation" taking place in an opportunity for value growth.

To grasp the potential arising from digital transformation, companies will have to change and be able to interpret new trends. Successful companies should be ready to reconsider their views and revise their traditional business model.

With this in mind, TIM launched a new model of relationship with leading universities and national and international research centers in order to enhance talent and develop key competencies to transfer innovation into the company.

The goal is to strengthen our ability to innovate and, at the same time, contribute to the development of research and training among young people together with schools and universities thus helping to bridge the gap between skills required by the labor market and those provided by education.

Our partnerships with academias, through the cogeneration of cutting-edge research, aim at forming the future ruling class, by giving support to students and giving a real economic value to knowledge.

Our collaborations consist in funding scholarships for PhD/merit awards/internships, exchange of lectures, sponsorship of Chairs on topics of interest for the company, sponsorship of codesigned Master programs, orientation and contributions to the design of educational programs in line with the skills required by the job market.

In a nutshell, we need new skills to enable our strategies and support change. Tim Academy is designed to satisfy this need by capturing, managing and disseminating knowledge within our company and transitioning to new contents through knowledge management.

Knowledge Management is represented by the set of tools and processes defined to build learning opportunities for people and for the company, accessible through specific tools and platforms of social collaboration.

We have defined a model and a process of Knowledge Management where knowledge spreads throughout the organization, thus making strategic knowledge available for business.

It is a model which has two fundamental characteristics: being cross-functional and innovative. In fact it is based on the involvement of multiple business functions which together generate new skills and ideas.

- Emersion and dissemination of knowledge, both tacit and explicit

The knowledge management learning model allows us to classify each of our initiatives in terms of focus and temporal horizon (see Fig. 13.2).

Furthermore, we have defined a Knowledge Management governance model which regulates the interaction of three teams:

- Strategic cross-functional team, which has the task of defining the objectives and the evolution of the knowledge management model in line with the business strategy;
- a Technological Development team aimed at implementing technological solutions;
- a support team dedicated to all activities related to creation, promotion, and communication.

The design and management of KM initiatives follows a standard consisting of the following phases: definition of strategic concept, technological solutions and governance, implementation and launch, ongoing activity and assessment & refinement.

Experiences and "lessons learned" that we are maturing and consolidating will be used to improve the implementation of the model and knowledge management initiatives.

In particular, among the Communities which are opened, we find one called Conferences Knowledge Sharing, which aims at collecting and sharing

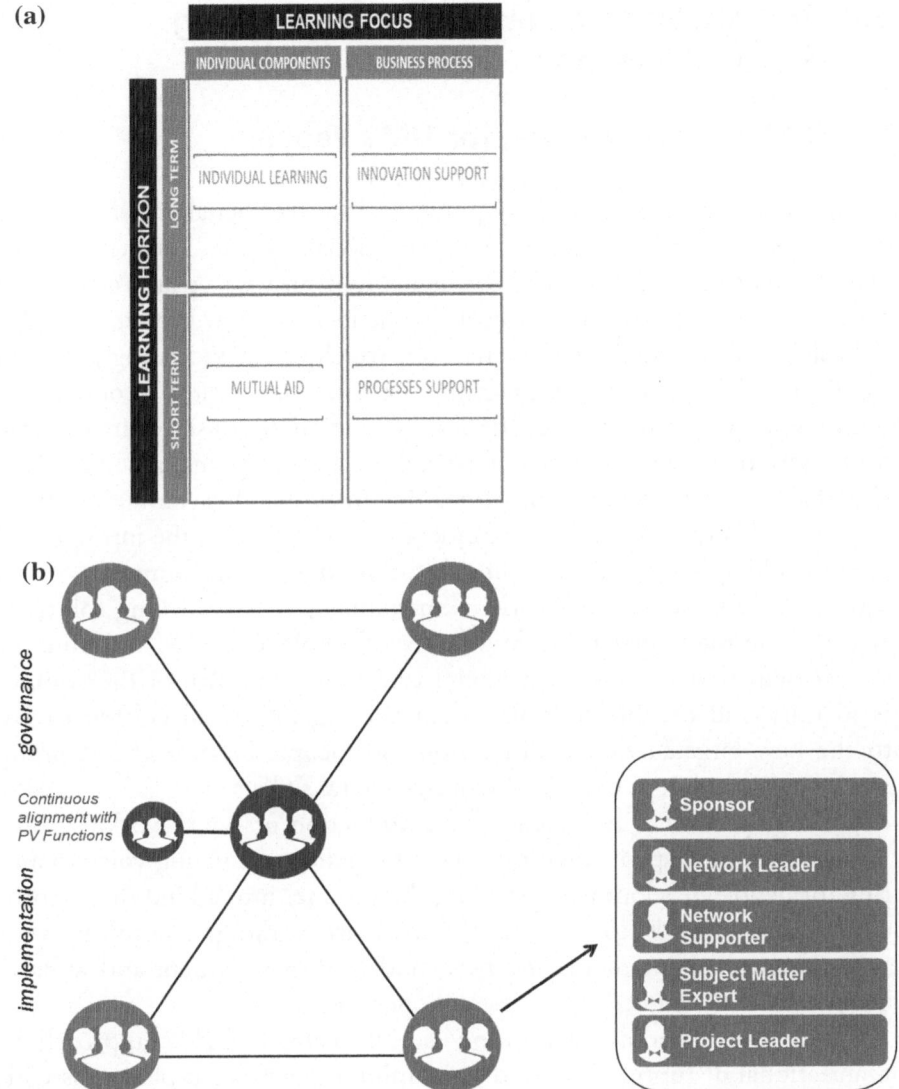

Fig. 13.2 Knowledge management model

knowledge gained by colleagues who participate in external conferences or workshops, other than defining a six month participation plan. Access to the Community is open to the entire company and also provides guidelines for the collection and publication of documentation and support for the organization of main highlights presented.

13.4 Transition to a New Way of Working: Smart Working[9]

13.4.1 TIM's Smart Working for TIM's People

Transition to new ways of working is induced by the digital transformation, which is characterized by decrease in transactional costs and increased flexibility, because supply and demand for digital jobs meet on "platforms" which compete on service prices (thus, outside of the logic of wage bargaining). This trend will also affect "traditional" work forms which will be increasingly focused on individual flexibility and autonomy in exchange for greater accountability toward results (which will replace the mere execution of tasks). Smart working has precisely these characteristics of flexibility, autonomy and accountability, and probably, for this reason, arouses growing interest and is coming out from the domain of corporate welfare instruments. In other words, the introduction of smart working is one of the enabling factors for digital transformation.

In such a scenario, in 2016 TIM is introducing smart working solutions which include the largest possible number of employees. TIM is aiming to create favorable conditions for a digital transition respectful of the company's identity and capable of making current organizational culture evolve into the new targeted digital dimension. So, people become the stepping stone for this transition, because it requires digital skills.

Consistently, TIM is leveraging the introduction of smart working solutions as a start to shape its own new digital organization giving voice to and caring for people development, evolving the intranet into a kind of platform for collaborative work, to ease (thanks to smart working) identification of new organizational paradigms and new models of management and welfare, to keep constantly aligned strategies and operations.

As a company whose organization is still analogic, TIM's approach to "organizational disruption" is based on combining new work-platforms with standard jobs without slowing the pace of change and bringing into the new digital environment those employees capable to adapting to it.

TIM's experience demonstrates that is possible achieve structural and social adjustments without slowing innovation if you engage your people to change. And smart working can generate engagement because of its win-win logic.

So TIM's Smart Working is neither a mere tactical move to gain some labor cost reduction applying for the fiscal incentives provided by Italian law, nor just a welfare choice to improve work-life balance, it is the leverage to achieve new, incremental productivity, both at "people" and "process" level.

From this last standpoint smart working is a «perpetual beta» change management journey, capable of improving productivity by increasing welfare and accountability for results (Fig. 13.3).

Along such a journey TIM deploys the three leverages of a typical Smart Working Project, those commonly known as Bricks Bits and Behaviors, enriched by two more company specific enablers, these are Business (going in alignment with the service offered to Customers) and Social (engaging People by empowering them to contribute and cooperate).

How the Business enabler works will be shown in the last part of this chapter, the following lines describe the Social dimension.

TIM assumes that ability to cooperate will be a must for the new ways of working brought by digital transformation; thus TIM pays special attention to the social dimension of working because considers it the cultural environment to develop the ability to cooperate. With this in mind, since the very beginning of the deployment of the Smart Working Project (May 2015), the company's intranet had a section on the matter. The early mission of this section was "communicate and engage". The development of the Project is transforming it in a platform for cooperating. On this platform information about Smart Working is available, and also is possible to apply directly to the smart working pilot with no need for a hierarchical intermediation, furthermore, people can support people having troubles with smart working technologies and can give feedback on Project's deployment and development.

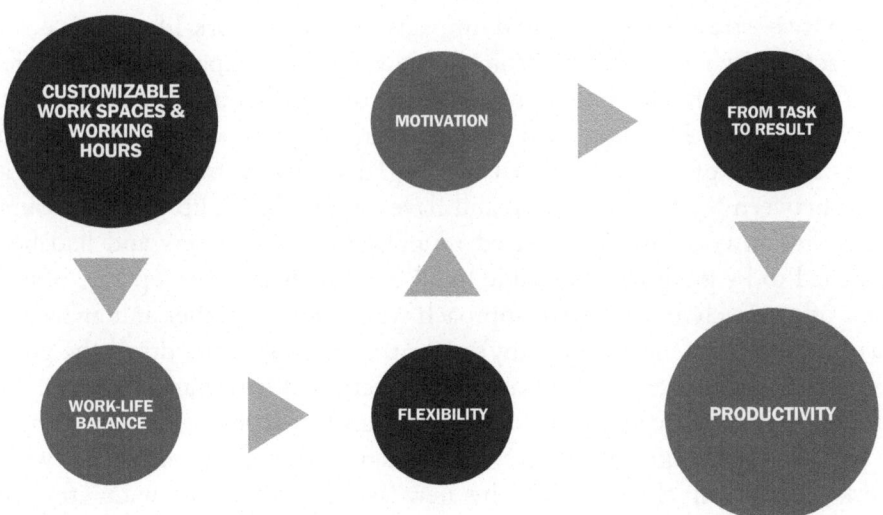

Fig. 13.3 Smart working as a "perpetual beta" change management journey

Such a company's intranet platform role and the parallel shift of personal equipment from plain tools to a system of services aiming to create a kind of personal technology platform for personal productivity are the whole realm of digital technologies inside TIM's approach to smart working, that is hinging on the transformation of individual behaviors both of managers and coworkers. To ignite the change of their attitudes TIM is working on behavioral training and on direct experience through different pilots gradually increasing how and how many people are involved by the smart working approach.

It is worth describing the solution adopted for training because it is consistent with the general purpose of digital transformation. Firstly TIM selected which skills are required for smart working and then clustered them into those for managers and those for coworkers. The result was that knowledge sharing, virtual communication, digital awareness, creativity, social collaboration, performance management, enterprise contribution are valid for both, while just empowerment and leadership styles for managers and self-empowerment and engagement for coworkers mark the difference between the two kinds of players. Then, TIM designed two training courses with these contents, which have been run in digital classrooms instead of physical classrooms. These lessons have been "taped" in order to be available on the training platform of the company's academy (TIM Academy) allowing employees to actively organize their own training on this subject.

Coming to direct experiences, the first move was (on September 2015) to run an extensive survey involving more than 30% of TIM's employees about their current ways of working; then (on October and December 2015) two stress tests measured impacts on both work-life balance and productivity when you work from home or from company's offices different from yours; finally, based on the results of October and December tests, on March 2016 TIM launched an agile working pilot with special working hours set to improve work-life balance (e.g. if you work from home, you can work between 8 a.m. and 8 p.m. and have a lunch-break up to three hours). This pilot in two months engaged roughly 8500 volunteers and had been designed to be gradually accessible to all employees that can operate outside their office. Such a progressive approach will generate, gather and analyze all adaptive reactions inside company's organization so to write down the policy for smart working capable to satisfy expectations for welfare improvement, productivity increase and processes digital transformation at the same time.

By Project design the largest transformation will be achieved where work spaces will change too. The new offices will break with "framed" spaces, cubicles that become a status symbol by dimension and furniture. New offices will be open plans of collaboration, real physical platforms for knowledge sharing where "solo" tasks and cooperative tasks will be both performable making possible moving from "Personnel" to "personal contribution". To achieve this condition TIM has a plan that probably is going to change the way the word "office" is intended in Italy.

Furthermore, such a sharp focus on digital transformation makes TIM able to choose new technologies and services by criteria consistent with the transformation for itself avoiding solutions that are not up to the target or beyond it.

The same sharp focus enables TIM to align the content of its digitalization with the solutions for smart working offered to Customers.

13.4.2 TIM's Smart Working for TIM's Customers

In Italy smart working or, as we call it, agile working is being a hot topic of organizational development since 2014. Companies' curiosity is evolving to benchmarking and quest for suitable applications based on digital communication and collaboration.

A specific market segment is born. In it because benefits are not granted by the mere adoption of technology, but are driven by behaviors change management, prospect customers prefer partners with real expertise achieved throughout direct experience to plain purveyors.

Then in this market you succeed if you can be a "role model". So for TIM adopting agile working is a competitive advantage because makes TIM the customer's partner of choice for digital transformation.

TIM by its approach to smart working is making a virtuous transition to digitalization. In this transition phase TIM can improve its smart working services by a "use what you sell" attitude and at the same time gains a positive visibility on media as an enabler of Italy's modernization.

So TIM's organizational development choice for smart working is a three-fold hit impacting on productivity factors, contributing to a better liveability in cities, being a B2B solution for digital transformation in a somehow "blue ocean" Italian market segment.

13.5 Conclusion

In this chapter, we discussed digital transformation with regard to its implications for organizations and the future of work. First, we described digital transformation analysing its drivers, both social and technological; we thereby focused on the profile of the new, "smart organization" and on the related skills called by this transformation. Then, we explained TIM's way, as to 2016, to enact the transition from knowledge to skills, through the TIM Academy initiative and the Knowledge Management process, and toward a new way of working, through the Smart Working project.

In TIM we think that digital transformation can be a serious opportunity to achieve future wealth for the company as a whole if it will be managed throughout people's skills and engagement.

Given the magnitudo of the desired transformation we are concentrating our effort on developing our capability to nurture new knowledge and ability to cooperate for innovation.

Along this journey we expect ourselves to be successful in digitally transforming not only our existing processes, but, as always, also in implementing a vision capable to enable future transformation thus allowing us to go beyond all our stakeholders' expectation.

Notes

1. Bruni, A. et al. (2013).
2. Orlikowski, W.J. et al. (2008).
3. Forschungsunion, Acatech (2013).
4. Global Challenge Insight Report—World Economic Forum (2016).
5. Drake, J. (2000). As to Italy, see also *49° Rapporto sulla situazione sociale del paese*, CENSIS (2015).
6. Rifkin, T. (2000).
7. Zammuto, R. et al. (2007).
8. De Stefano, V. (2000).
9. A variant of this paragraph was published by Andrea Iapichino as "L'esperienza del lavoro agile in TIM" in "People management", edited by Mauro Tomè, S. Deiana, D. Patruno, and L. Redaelli. Wolters Kluwer, 2017, Milan, Italy, p. 211.

References

Bruni, A., T. Pinch, and C. Schubert. 2013. "Technologically Dense Environments: What For? What Next?" *Tecnoscienza. Italian Journal of Science & Technology Studies* 4: 51–72.
De Stefano, V. 2000. The Rise of the Just-in-Time Workforce. On Demand Work, Crowdwork and Labour Protection in the "Gig Economy", International Labour

Office, Conditions of Work and Employment Series, 72, 2016.d-for Experience, J.P. Tarcher/Putnam, London.

Drake J. 2000. *Downshifting: How to Work Less and Enjoy Life More*. San Francisco: Berrett-Koehler. As to Italy, see also 49° Rapporto sulla situazione sociale del paese, CENSIS, 2015.

Forschungsunion, Acatech. 2013. Securing the Future of German Manufacturing Industry. Recommendations for Implementing the Strategic Initiative INDUSTRIE 4.0. Final Report of the Industrie 4.0 Working Group.

Orlikowski, W.J., and S.V. Scott. 2008. "Sociomateriality: Challenging the Separation of Technology, Work and Organization." *Academy of Management Annals* 2: 433–474. https://doi.org/10.1080/19416520802211644.

Rifkin, J. 2000. *The Age of Access: The New Culture of Hypercapitalism, Where All of Life is a Paid-for Experience*. New York: Tarcher-Putnam Books.

World Economic Forum. 2016. The Future of Jobs. Employment, Skills and Workforce Strategy for the Fourth Industrial Revolution, Global Challenge Insight Report, January, pp. 5–8.

Zammuto, R. et. al. 2007. "Information Technology and the Changing Fabric of Organization." *Organization Science* 18 (5): 749–762.

14

Investigating the Potential for Micro-work and Online-Freelancing in Sri Lanka

Helani Galpaya, Suthaharan Perampalam
and Laleema Senanayake

14.1 Introduction: Outsourcing and Micro-work in Sri Lanka

14.1.1 Sri Lanka as an Outsourced Work Destination

Work transformations are being driven by globalization and technological revolution and there is a shift happening from mechanical to digital technology (UNDP 2015, chap. 3). The phenomena of technology changing the boundaries of the firm is not new, and has been documented since the first half of the twentieth century (Coase 1937). The first wave of business process outsourcing took place thanks to the advent of high-speed digital connectivity, which enabled the transfer of vast amounts of data across to locations that had lower cost or other advantages. At first work was transferred from one firm to the same firm's subsidiary located in the cheaper overseas location. Over time, work started being transferred to non-captive firms.

This research was carried out with the aid of a grant from the International Development Research Centre Canada, and the Department for International Development (DFID) UK.

H. Galpaya (✉) · L. Senanayake
LIRNEasia, Colombo, Sri Lanka

S. Perampalam
LIRNEasia, Colombo, Sri Lanka

© The Author(s) 2018
L. Pupillo et al. (eds.), *Digitized Labor*, https://doi.org/10.1007/978-3-319-78420-5_14

Sri Lanka is a lower middle-income country that has benefited significantly from the increase in business process management (BPM) work. Many global BPM operations have set up business in Sri Lanka and provide a range of services to overseas clients, ranging from the low-value added out-bound marketing calls or data entry, to in-and-out bound customer service centers for global banks, to high-value added activities such as medical image processing (reading x-rays), financial account processing and equity research for investment firms. The country has been a desirable location for offshoring for over decade—ranked in the top 10 in Asia (WEF 2015), 14th globally (A.T. Kearney 2016), or 16th globally (Tholons 2016). The BPM sector is projected have strong growth rates for next few years (A.T. Kearney 2015).

Over a decade ago Sri Lanka's private sector and government realized that the country had one of the highest per capita rates of qualified charted accountants in any country. It was also leader in the Human Development Index, indicating readiness in education and health attainment. Therefore, instead of only pursuing the inbound/outbound call center market (which is extremely price sensitive, difficult to attract and retain talent in, and dominated by India and the Philippines), an active push was made toward higher-value work such as accounting, accounts processing. This, combined with improved high-speed Internet connectivity, lower labor costs, implied continued growth in the higher-end of the BPM sector.

The Information technology (IT) industry, (comprising of software services and software product firms) is also booming, with high profile acquisitions of Sri Lankan software firms by global players such as the London Stock Exchange.

Together, the IT-BPM industry has over 300 registered firms (DailyFT 2014), earned over USD 850 million in 2015 (SLASSCOM 2016) and employed over 82,850 people (SLACC 2016). Employment in the sector grew by 17% between 2003 and 2010 (Department of Census and Statistics 2015a). The sector contributed 0.15% to the Gross National Product (GNP) in 2016 (Central Bank 2015).[1]

14.1.2   Online Freelancing/Micro-work Platforms

The past two decades has seen another type of outsourcing, namely, freelancing and micro-work. Micro-work breaks down large chunks of work into small and simple tasks that rely on human intelligence and distribute these "micro tasks" to workers via Internet for greater cost efficiencies across geographic boundaries (Kuek et al. 2015). Online freelancing refers to slightly large chunks of work which are outsourced online, but don't involve crowd-sourcing or the automated "assembly" of many micro-tasks to complete the whole (see Hoßfeld 2011 for a typology of online work and differentiating characteristics).

Unlike the traditional BPM industry that uses bilateral contracts between two firms, micro-work outsourced work does not involve long-drawn out/ negotiated bilateral contracts. Often, each job is of extremely low value, enabling the buyer to outsource the work to more than one seller, paying for (i.e. "buying") all jobs that are completed, but only using the best output. The seller of services is usually an individual, and often the buyer is too (as opposed to firms/institutions). Commonly found "jobs" or "gigs" on micro-work platforms include ad-clicking, media tagging, data input, transcribing, data verification, information gathering and summarising, proof-reading, translation, copy editing, graphic design, website design, and website de-bugging.

The above is done online via platforms which allow for end-to-end trans-actions to be carried out. Starting from market discovery (where the buyers and sellers announce themselves/their work and find each other), contract-ing (the buyer sources the work from the seller of service, agree on price if applicable), service implementation (the buyer and seller communicate dur-ing the carrying out of the work, the seller completes the work) and service/ produce delivery (the seller delivers the out to buyer), payment and post-sale feedback/review of quality. Workers (sellers) advertise themselves by adding a profile on the platform. The profile could include a listing of skills, upload-ing/links to previous work. Buyers advertise the jobs. Buyers can select the worker directly. The selection may include the buyer examining the sellers previous work and his/her reputational rank on the platform. Alternately, depending on the platform, the buyer can advertise the job and ask sellers to bid for the job, then select one based on lowest price or other criteria. On most websites, the sellers are ranked based on multiple criteria includ-ing a satisfaction ranking assigned by past buyers as well as other indicators automatically measured by the platform itself (such as response time to buyers' communications). Payments are done via the platform, which charges a commission (percentage of the value) on what the seller is paid.

The traditional BPM industry requires the provider of services to main-tain provide quality assurance through a service level agreement. Micro-work platforms usually don't take on a explicit quality-assurance role. But they do facilitate quality signaling of sellers by allowing buyers to rank sellers based on work done, and allowing sellers to take standardized tests/exams on the platform itself to prove various skills.

For workers (sellers), online outsourcing has generated new opportunities to access work in a global market, anywhere at anytime, as long as they have a computer, Internet access and the required skills (Kuek et al. 2015).

14.1.3   Micro-work in Sri Lanka: Is There Potential?

As Table 14.1 shows, there is awareness of micro-work in Sri Lanka—three specific platforms (Fiverr.com, Freelancer.com, Upwork.com) are popular among web users, and very popular in Sri Lanka relative their popularity in the rest of the world. A relatively high number of workers are registered as potential service providers.

The standard entry-level qualification on recruitment in the IT-BPM this sector was a Bachelors degree, with 63% of the workforce holding a graduate or post-graduate level qualification (Ph.D., Masters Degree, Bachelor's Degree, Post-Graduate Diploma) in 2013 (ICTA 2013). This means that the employees in IT-BPM sector are the educated—employees include English speaking graduates who have qualified from the highly-selective local universities, branches of overseas universities present in Sri Lanka (attracting those who can afford the fees), universities overseas (which are even more expensive, therefore presenting limited opportunities for the masses), and private technical degree granting institutions which are not classified as universities. Sri Lanka produces around 25,000 university level bachelor's degrees per year, and only 5,778 are computer science and engineering or related topics. They are traditional feeders to the IT-BPM, telecom services and related sector jobs.

Table 14.1 Proxies for the use and popularity of various micro-work platforms in/by Sri Lanka (*Source* Authors, based on http://www.alexa.com/ and websites of the listed micro-work platforms. The Alexa ranks websites by frequency of access, by country. Registered number of participants is a count of the sellers who self-declare Sri Lanka as their country when registering on each platform)

Micro-work platform	Site rank by Alexa		Registered number of Sri Lankans
	Sri Lanka	World	
Upwork.com	264	579	5000
Freelancer.com	289	1424	5003
Fiverr.com	67	526	Fiverr does not allow sorting registered sellers by country. But the number likely to be higher than Upwork/ Freelancer given the very high Alexa ranking in the country
Microworkers.com	500<	17,905	–

But outside of this, there are persons with skill-sets that are marketable via online platforms—for example, around 1250 men and women complete professional courses that fall short of a bachelor's degree (Gamage and Wijesooriya 2012) Thousands more partially complete diploma courses related to computer literacy, English or graphic design.

Sri Lanka has the highest literacy rate in South Asia, at 92.5% in 2012 (Central Bank of Sri Lanka 2015). 36.5% of the total population was able to read and write in English with 23.8% able to speak it (Department of Census and Statistics 2012). 26.8% of countries population was computer literate[2] in year 2015 and over 25% of households had a computer (Department of Census and Statistics 2015a, b). 19.5% of the population had Internet access (Central Bank of Sri Lanka 2015). While there is much room for improvement in computer literacy and Internet access, there is broad enough diffusion as to make participation in micro-work as service providers a viable option for many, beyond the elite and educated.

Youth unemployment (at 20%) is significantly higher than the national unemployment rate (of 4.6%); female unemployment (at 7.6%) is also higher than average unemployment (Department of Census and Statistics 2015a). Finding information technology related work for these groups is a possible solution to the unemployment problem.

Therefore, **is it possible that the digital dividends could be spread more inclusively (beyond the educated elite) through the participation on micro-work platforms? Is this already happening? If not, why not? And If yes, how do we encourage it?**

These are important questions for policy makers in a developing country. We therefore attempted answer the following research questions using a mix of methodologies.

- What is the current incidence of awareness, and participation (i.e. doing work) on micro-work platforms in Sri Lanka?
- What are the skills needed to do the type of jobs commonly available via popular micro-work platforms? What is the availability of such skills in the country?
- What factors make micro-work an attractive (or unattractive) employment option for people with the requisite skills?
- What are the barriers (beyond attitudes) faced by those working on such platforms, or those hoping to work on such platforms?

14.2 Methodology

To answer the above questions, we conducted two separate but connected lines of research using two methodologies. To understand available skill levels in the country, we inserted questions into an ongoing nationally representative sample survey of media use. To understand attitudes toward and experience of micro-work, we conducted a series of focus group discussions. For both, our target age group was the population between the ages of 16–40. The lower level 16 was used is the age when students sit for the Ordinary Level exam (after 10 years of schooling), after which some go into the workforce. Existing literature suggest majority of the online freelance workers are young and between the age group of 18–28 (Kuek et al. 2015). But we wanted to understand the dynamics of such work among slightly older persons who might have the necessary skills to participate on platforms. Therefore our upper age cut-off was set at 40.

14.2.1 Quantitative Sample Survey

Quantitative findings of this paper are based on a nationally representative survey of Sri Lankans aged 16–40. The sample size was 5500 and as designed to represent the target population, covering both urban and rural areas in all provinces and districts of Sri Lanka with $\pm 2.5\%$ margin of error. The respondents were selected using a multi-stage stratified random sampling method using probability proportional to size (PPS). Fieldwork for the study was conducted in October–December of 2015. The structured questionnaires were designed in English, translated to the local languages (Sinhala and Tamil) and field tested and implemented in the same.

14.2.2 Qualitative Research Protocols

The qualitative research was designed to understand perceptions and attitudes towards online micro-work. Six focus group discussions (FGDs) and one in-depth interview (IDI) were carried out in the three population centers: Colombo (4), Jaffna (1) and Galle (1). The discussions were conducted between two separate groups: (a) potential workers (those who have some of the basic skills such as a diploma in computer science, and might be potential candidates for working on online platforms), and, (b) those who are currently engaged in online micro-work. All participants were between the ages of 16–40. Each FGD had between 3 and 6 respondents, and lasted

2.5 h on average. The IDI was with a respondent who could not participate in a FGD, and lasted 2 hours. Of the 30 respondents 18 were male, 12 female; 17 were current micro-workers while 13 had skills but had not done micro-work. Education qualifications of respondents ranged from those with 12 years of education to those with bachelors or equivalent degrees. The protocols were conducted in the two local languages (Sinhala and English) by the authors between February and April 2016, using a semi-structured questionnaire. The conversations were recorded with participants' consent. The recordings were translated to English and transcribed. The authors then analyzed the English transcripts.

Given there is no way to personally identify or contact micro workers through platforms (only their online profile/username is available, not their emails or phone numbers), and given there is no existing listing of such workers to sample from, the research term used a combination of methods to recruit participants for focus groups, including (a) posting a "job" on the platforms, promising payment for workers who fit the screening criteria and agreed to participate in the research, (b) using known persons who worked on such platforms to identify others who did the same (snowballing), (c) using the services of a market research firm to recruit from the field, in return for payment, (d) attending "micro work training programs" provided by third parties in the country, and identifying potential recruits who attend such events.

14.3 Results and Discussion

14.3.1 Low Awareness of and Low Willingness to Do Online Freelancing/Micro-work. Those Willing Are Only Interested in Doing So on a Part-Time Basis

The national sample survey results show that about quarter of the target population are aware of online freelancing or micro-work. Males are significantly more aware of such opportunities compared to females. The younger and richer the respondent, the higher the awareness. We hypothesized that rural areas may have little awareness of such opportunities, but this turned out not to be the case: 25% of the rural target population are aware of such opportunities, a number higher than we anticipated, compared 31% of those in urban areas (Table 14.2).

Table 14.2 Awareness of and willingness to work online among Sri Lankans aged 15–40 years (*Source* Authors, based on data from nationally representative sample survey)

	Awareness on micro-work/ freelancing		Willingness to work on micro-work/online free-lancing jobs		Base/all respondents N
	Aware of online work-ing opportu-nities (%)	Not aware about online work oppor-tunities (%)	Willing to do micro-work (%)	Not ready to do micro-work (%)	
All Sri Lanka	26	74	11	89	5377
By gender					
Males	32	68	14	86	2227
Females	21	79	8	92	3150
By area					
Urban	31	69	10	90	2633
Rural	25	75	11	89	2326
By SEC					
SEC A and B	43	57	16	84	1465
SEC C	27	73	11	89	1626
SEC D and E	15	85	8	92	2025
By age group					
16–23 years	35	65	19	81	1199
24–31 years	25	76	10	90	1661
31–40 years	19	81	5	95	2517

After the initial questions about awareness, the enumerators explained what is meant by micro-work/online-freelancing and gave examples of work available through online platforms. This was followed by questions about whether the respondent is likely to get involved in such work. Only 9% of the target population responded positively. 12% of males were more amenable to such work, compared to just over half as much (7%) for females. Those living in urban areas showed least willingness, possibly reflecting the availability of other employment opportunities. Rural residents were more keen on micro-working/online free-lancing. But disproving our hypothesis that only those without access to other job opportunities would be attracted to such work, when we dis-aggregated the data by income, those from the richest households (Socio-economic classification (SEC A and B)) were more willing than those from households with lower SECs. Yet this could also be a reflection of the luxury the rich have in being able to undertake non-permanent, "gig" work, compared to the less affluent respondents who prioritize regular income/salary due to financial constraints. The youngest group was more keen than the older groups to try such work.

We dis-aggregated the "willingness to do micro-work/online-freelancing" by current employment status and skill level of the respondent. Surprisingly, the employment status didn't seem to change willingness to do this type of work, with only a small difference seen in the responses of the two groups. But self-assessed skills of the respondent clearly had an impact. We looked at all those with the minimum skills (including basic English, computer typing and data entry) required to do even the lowest-value added work on platforms (such as ad-clicking or entering data). In this group, those that had higher confidence in their skills were more likely to express willingness to do micro-work/online-freelancing once the concept was explained to them (Table 14.3).

Among respondents who *are* willing do this work, there was a significant preference for doing so as a part time job. This was true among both employed as well as the unemployed. More women were interested in part time work online compared to men (Table 14.4).

The quantitative study revealed the low awareness and low willingness to do online work of this type. Even when people were willing, they were only willing to consider doing so on a part time basis. And to our surprise, willingness wasn't radically different between the employed vs. the unemployed; the only meaningful difference was between men and women. The only "expected" outcome was that people with higher skill levels (among skilled required to do the popular jobs offered on the platforms) being more willing to consider online work.

Why do we see these patterns? Is it that irrespective of employment status, there are such high negative perceptions about micro-work/online freelancing

Table 14.3 Willingness to do micro-work/online-freelancing based on employment status and skills levels of Sri Lankan 15–40 years (*Source* Authors, based on data from nationally representative sample survey)

	Willingness to work on micro-work/online freelancing jobs		Base (all respondents)
	Willing to do micro-work (%)	Not ready to do micro-work (%)	N
All Sri Lanka	11	89	5377
By employment status			
Employed	10	90	2167
Unemployed	12	88	3210
By skills level			
I do not have any skills	4	94	3967
My skills are basic and I did not get any training	18	82	685
My skills are basic and I got some training	31	69	556
My skills are excellent and I got advance training	53	47	170

Table 14.4 Preference for full time vs. part-time micro-work/online freelancing (among Sri Lankans 15–40 years who are willing to do such work) (*Source* Authors, based on data from nationally representative sample survey)

	Nature of commitment		Base/those who are willing to do micro-work N*
	Like to do part-time (Less than 40 Hrs.) (%)	Like to do Full time (More than 40 Hrs.) (%)	
All Sri Lanka	77	23	594
By gender			
Males	75	25	379
Females	82	18	215
Working status			
Employed	75	25	216
Unemployed	81	19	378

that most people won't do it Is it because other "traditional" work is more valued in some way? In our qualitative research (focus groups) we paid particular attention to such questions. Sections below highlight findings from the qualitative research, with specific/direct quotes (from the English transcripts) given where possible.

14.3.2 Awareness Came Through Multiple Channels

Online ads (including Facebook ads), blogs, newspaper ads and friends/family were the ways in which first awareness was created.

> *I got to know about it from a blog post. It had step by step, how everything has to be done. This blog post was done by a Sri Lankan.*

> *One of our lecturers* [at the computer training institution] *recommended Fiverr as a work platform.*

> *One of my brothers was working in Upwork. I got to know from him.*

Several firms also advertised in newspapers that for a fee (of around USD 30 or LKR 5000) they can teach people how to "make money from home" by doing online work. Such firms mostly exposed new workers to online ad-clicking or similar low-value-added work. Some of these training institutions acted as workers on the platforms themselves, hiring newly-trained staff as "subcontractors."

What they do is, they bring people and train them. They have an account. Odesk or somewhere. They register you as employees of their business. They take work from others and give them. And he submits the work to the buyer. He takes a large amount from the buyer and distributes the employees only a small amounts

14.3.3   Getting a Foothold on Platforms Is Not Easy, and Signaling Quality Is Important

Most sites have a rating system for sellers—higher ratings being a strong signaling mechanism for future buyers. Sellers employ various strategies to improve their ranking.

When a job for USD 55 was posted, some people ask if the 55 USD can be divided into 5 USD tasks. That means its 11 tasks. Some people ask whether we can add 11 orders.....From the same person, but he gets 11 reviews...[and goes up in ranking]

Some platforms (for example UpWork and Freelancer) offer exams that the sellers can take. Passing these exams is another signal of quality. Such signals appeared to be particularly important for those who had not done work on the platform before, and had no previous references or ratings.

They [the website] asked me the projects that I've engaged before. I don't have previous experience. There are online exams. When I do these exams the profile becomes 100% complete and we can work. So I did these work. Then they let me work on the platform.... the exams are free.... If its English, they have English exams. For Word, there are Word exams. You should sit for the exams that you have to work in

These quality signals in turn have a huge impact on how often a seller gets work.

All the hard work is up to the point where you build your profile in freelancing platforms. Once you developed the profile and reach certain level of ranking you can easily make money being at home. Today I am getting got lot of orders from specially from Europeans and Australians. Now I am able to quote more [per job], since I developed a good name and commitment with my clients. With the client base I have co-founded my own venture and it is running successfully.

Getting ranked therefore depends on how many previous jobs one has done, how satisfied the buyers were, how responsive one was on the platform when dealing with a buyer and host of other factors. But to get ranked, one has to get on the platform and at least get one job. This first job is the biggest barrier—no one wants to buy services from someone who has never done a job. Successful workers overcome this by "gaming" the system, using their contacts/friends, or by offering to do the (first) job for free. They learn such tactics from their friends who have done the same in the past.

> When I put [a profile] up, a customer messaged me. The first job, I did it at a loss....I told him, if he buys a gig, I'll give him another one for free. This was because I wanted to get orders. So, I kept on working with him for a month. After that I kept getting [paid] orders.

> I did a trick, asked my friend to purchase several gigs from me and give me good ratings and I paid him back. This is how I built my profile before getting the first job.

Many who never crossed this "first job" barrier never got any work despite registering.

> I registered with most of platforms (Fiverr, Freelancer, Upwork), I am maintaining profile each of those platform. I bid for all type of logo designing projects. Competition is very high in online freelance platforms, you are competing with global work force to get the same job....I still apply for some projects in Fiverr but I am not successful, may be I also have to apply some tricks.

14.3.4 Flexibility Offered by Freelancing/Micro-work Platforms Is Attractive

Majority of the current micro-workers we interviewed work on the platforms on a part-time basis, while engaged in other full time jobs, or while studying. They were able to take up the micro-work due to the flexibility.

> Most of the time I work at night. When I get an order, I should deliver it within 2 days. So when I get an order, if today is a working day [in my regular job], I go home and finish it today itself. I don't have a specific time per say. I work whenever I can

> I go to university and come and do this work during evenings

The choice of different types of jobs on these platforms meant workers can select the jobs that fit their time availability—working intermittently or at a stretch.

If I do a brochure or leaflet, I can only do one job during this time. The time it consumes is high, it takes about 2,3, hours. But we can do 2, 3 visiting cards within an hour. So, I apply more for these type of small tasks

Respondents also select how much they want to work based on their income needs.

I don't run after this. I earn like 10,000 LKR per month. I try to cover my basic expenses. If I think that I want more money I work or else I will just stay [without working]

14.3.5 The Range of Work Is Varied. Higher Value Jobs Enable Workers More Control Over Price

Many with basic computer skills start at the low-value-added end of the spectrum which consists of ad-clicking or entering the data seen on images on screen. These workers have no negotiating power (over price).

Numbers come up in the site, and we have to look at them and click. It shows a number for a certain time. We have to look at the advertisement they stream, and they display a number and tell us to click on that number. When we do that, USD 1 is added [to our account].

[I do] Data entry. They normally send me scanned documents. I look at those things and type.

Social media marketing also repeatedly showed up in the type of jobs many respondents would carry out. Such work gave slightly more opportunities for the worker to make/negotiate price offers.

This guy wanted me to write content and also to publicize one of his tournaments as much as possible to as many people as possible. So I wrote content, and I told him, ok I deliver this to over 25,000 people for this price.

Higher value work was done by those with software development, software testing/debugging skills. These higher value jobs were referred to as "projects."

I liked to work project based. Because I normally do web development. What I do is I fix issues in small business customers. I don't do full projects for 5- 10 USD

In platforms that matched higher skilled workers with jobs, it was normal for the workers to bid/name their price.

> *In Elance, Odesk, you go on the platform, look at the project and the scope, and if it can be done we say OK and submit saying this is the cost, this is the proposal and we can deliver at this time. Here, we have control.*

> *[How much I earn per project] depends. As an example, for a small taxi service, think it's a airport drop, people put up a job to create [the website for] an order, pick up venue, time, drop time, etc. So to create this, I quote the project for 400 USD.*

14.3.6 Perceived and Real Barriers Make Micro-work/ Online-Freelancing an Unattractive Full-Time Work Options

Rarely did we hear from respondents about people quitting their regular/full time jobs and doing online freelancing/micro-work full time. The majority did work on these platforms as a part time activity, in addition to their full time job. Others were students who considered freelancing as a transitory activity.

The security of income of full-time work was attractive to many. Uncertainity of income and uncertainties due to technology challenges were cited as risks of online work. This was the case even when they had consistently earned more through online work compared to their regular/full time job.

> *Moderator: So if you can get more jobs in Fiverr, will you quit your full time job and work in Fiverr?*
> *Respondent: No I will not. Its because I'm not sure of the situation. Our accounts can be hacked, blocked etc. If something happens, I will loose my income. And there is no permanent income in these work*

Workers also said it was more profitable to do some jobs in the local market (face-to face for a local client, on a contract basis) instead of doing the same work online. They also felt that face-to-face customer relationships were stronger than those online.

> *....more than working on Fiverr, if I make a logo in the local market, I can earn more here than on Fiverr*

The competition [on Fiverr] is high, and there is a temporary bond between the customer and me. If I go to the local market as a freelancer, customers try us. So, the bond we create with the customer is high. So, we can earn more from this work.

Adding to this was the risk of not getting payment from these platforms.

The other risk of Fiverr or whatever, if something happens someday, if an issue arise, say its because Sri Lankan regulations, or because of some other reason, you might loose everything all of a sudden. If that happens, I'll loose my income. In this context, if I had a [traditional] job, with experience, I can go for another job.

For others, the lack of career path in online work made it a purely transitory earning source. For example, content writers did some of the more differentiated types of work. But even they didn't see a future in this work or found the work not challenging.

So, if you want it I think you can have it as a full time thing. For me, it's something on the side. I love writing, but I feel that writing about cookies, cup cakes and trucks it doesn't build up my writing. It doesn't challenge my writing skills.

One of the most significant challenges faced by online workers was the inability to prove their income when attempting to access formal financial services.

There is an issue for Freelancers. When we go to a bank, to get a loan, then ask whether we work in a place where EPF or ETF [government mandated retirement contribution schemes] is deducted. What they truly ask is, from where we receiving money. So, when we say that we are doing an online job, then the bank decides, "Oh, its not a stable job." Then they refuse to give us loans saying that it doesn't comply with their regulations. But this is not an issue in other countries. This is the most pressing issue that Freelancers have

Like most of Sri Lanka, the respondents used pre-paid data bundles for Internet access. Even though entry-level Internet data packages in Sri Lanka are among the lowest in the world, respondents felt they were paying too much for data packages which would reach the capped maximum in a few days. Some were unhappy with data quality, in particular data upload speeds (most ISPs focus on download quality).

Even though the speed is really fast, if the GB level is low, its of no use. It will finish in 2,3 days.

They [potential buyers] tell us to upload sample videos [before giving a job to create a video]. If we upload these samples, with the data that was used, the amount that we earn will be of no use.

We need uplink [speeds]. Most provide download speed but upload is low. If we host a website, what we need is upload. Its difficult to do it here

As one respondent summarized:

The greatest issue is Internet, next is payment methods, next is electricity, power failures.

Apart from these, cultural attitudes also played a role—lack of acceptance by family and friends discouraged men and women from doing micro-work. Though once earning ability is proven, there seems to be a higher level of acceptance.

They don't understand what I'm doing

…in Sri Lanka there is a culture that has been followed where we have to dress and go outside for a job.

For me of course at the beginning they [parents] kept telling me to do a [proper] job, don't do this type of work, later, when I started performing, they started supporting. People at home was afraid of what will happen because I'm not doing a day job

14.3.7 Cashing Out Earnings for Work Done Online Is a Problem

Many respondents (or someone they knew) had done work on certain platforms for an extended period of time, but were unable access the money they had earned. After the failure, many stopped online work altogether.

Then I tried it [ad clicking] out on the computer by my self. I did ad click for several years but I couldn't withdraw money

Some were directed by the micro-work site to open a Paypal to get paid. Sri Lanka doesn't allow Paypal payments. Therefore, even if the micro-work site transferred money to the worker's Paypal account, there was no way to access that money.

They [the micro-work site] told me to create a Paypal account. Which I did...I asked local banks such as Hatton National Bank and Commercial Bank. What they said is that it cannot be done. The money cannot be withdrawn from Sri Lanka, it can be only done via Singapore. So, then I gave up on the idea of withdrawing that money

Without clear information on how to claim money for work already done when cashing out via Paypal fails, workers blame both the micro-work platform and payment intermediary (which is Paypal in the example below).

That site [the micro-work platform] says that it added USD 1000 to the [my] Paypal account. But when I login to the Paypal account, it doesn't show that money. So, later when I found out, I got to know that, when I create a Paypal account and I add my country as Sri Lanka, what happens is, Paypal itself takes that money. ... That's what people said when I asked around

Some who had friends who had been through this experience before, or had heard about the payment problems in some way, limited themselves to the platforms that offered a clear payment solution. For example, Fiverr was popular because it had recently launched an online debit card.

This [reluctance to use other platforms] is because these sites are linked with Paypal accounts. There is no way in Sri Lanka to convert digital money to physical money.... Because Fiverr has a card, we can get physical money.

Specifically, the debit card offered by Fiverr enabled a worker to accrue earnings into the debit card account. Then the worker can go to an ATM machine in Sri lanka to withdraw cash, or use the debit card to make purchases at any location that accepted major credit cards.

They [Fiverr] give the card in partnership with Pioneer. I can load the money to this card. ...I can load the money that I get from Fiverr and cash out from a master ATM located here

This card is not cheap—the card issuer charges USD 3 fee per month and a fee is charged every time the worker draws out cash. Yet this card seemed to be only way to legitimately cash out the earnings. Nearly all current workers we talked to used the Pioneer card.

Other round-about ways existed to cash-outs the money that was earned—for example using a Paypal account that was registered in another country, or creating a Paypal account with a Sri Lankan address and using the accumulated money to purchase things online (without cashing out). But most respondents considered this very limiting, and preferred to obtain cash.

> … *the Paypal account that I use is in Singapore. It was done by a friend of mine who is in Singapore. Its because it cannot be done in Lanka…..[my friend] withdraws money in Singapore and sends to me*

A handful of tech savvy respondents had managed to open a "Malaysian" Paypal account from Sri Lanka by faking the IP address of their computer at the time of account opening. This enabled them to transfer money to their bank account in Sri Lanka.

14.4 Conclusions

We see from our research that while there are exceptions, for most, micro-work/online-freelancing is only a secondary source of income, something to be done in addition to a full time job, while waiting for a full time job or while in education. Traditional cultural preferences for a "office jobs" played a role here—it is less acceptable to say that one is "working from home," especially if one is male. However, once parents/family start seeing high incomes from such work, the negative perceptions appear to lessen. For many, the variable nature of the income was also deterrent, though it was unclear if the variable nature was because they engaged in micro-work part and never really had time to give 100% to developing a steady source of income online. Inability to prove income when attempting to access formal financial services was also a deterrent because local banks don't have a mechanism for assessing credit worthiness of workers who have variable income, or income from online sources that cannot provide employment references. The inability to get payment for work done due to Paypal not being legal in Sri Lanka was repeatedly cited as a barrier. We talked to workers who had done highly-commoditized jobs (ad-clicking, specifically) for almost a year until they accumulated the minimum amount mandated by platform before cashing-out was allowed. But when they attempted to cash out, they were unable to, because Paypal could not transfer funds to Sri Lanka. The Central Bank of Sri Lanka has been promising for years that this would be done, but changes to policy are yet forthcoming. Even more impactful would be for the banking sector to create account types or credit scoring mechanisms that enabled irregular income earners, self-employed workers, or informal sector workers to have access to formal financial services.

General lack of awareness (with less than 25% of the target population knowing what micro-work or online freelancing is) is a problem to expanding this market—without knowing of opportunities, few will look for them. More importantly, few will acquire the skills needed for online platforms. Everyone who gets the basic computer skills will expect to obtain jobs in the highly competitive software industry, or abandon those skills and find other employment, instead of realizing there is an intermediate market which values their skills.

The worker's level of skill is one of the biggest determinants of how much one earns. The highly commoditized end of the spectrum (ad-clicking) earns little income. The workers are price-takers and have no ability to differentiate themselves. In the middle of the spectrum, jobs such as logo design require more skill, but are still sold at platform-specified prices, is being commoditized and facing severe price competition due the entry of low cost workers from India, Pakistan etc. The higher-end of the market involves developing software/web functionality. Here, the workers are often able to negotiate a price with their buyer—if not on the first job, certainly in subsequent jobs. This type of work also often leads to repeated engagement between buyer and seller (with most taking the work off- platform and building a direct relationship). It is here that Sri Lanka needs to target—and promote this type of online work as a viable full-time or part-time option for those who cannot get employment in Sri Lanka's world-famous software firms, but still posses basic coding skills. This end of the spectrum is not necessarily safe from competition either, and is being undercut on price by Indian and Pakistani workers, but because of the longer term relationships, enables the sellers more control over time, price, and quality.

We see success-stories of workers earning significant income each month through online-free lancing and micro-work. We have even observed many who start at the low-value-added end (ad-clicking) or logo design, and move onto higher value jobs such as web-site design. The common factor in those who made this "migration" was peer learning. How to get paid on a platform, how to get the first job, how to get a higher ranking: these were all "tricks of the trade" that most learned from their network of peers who had done it before. Many who dropped off or never migrated to higher value platforms did so because they never learned these strategies. It is possible that more online or offline "meetups" and networking events can be organized to help develop peer networks for current and potential workers.

Such networks will help overcome real barriers as well as perceived barriers (which keep potential workers from getting online and seeking work).

Despite specific attempts to recruit females who were stay-at-home mothers and were also current micro-workers, we only found one respondent in our research (other women we interviewed were students or had other employment). This is at least some indication that there is a huge untapped market—women's participation in the labor force can be significantly improved (from the current low figure of 26%) through online platforms. By enabling women to work from home, those with young children will be able to achieve some work-life balance without stepping off the workforce all together.

Whether people engage in it as full time or part time work, online work presents a way to increase their income, sometimes significantly. Sufficient number of people with the right type of skills exist in the country. However, it has to be "pitched" right—awareness has to be increased, but only along with the pitfalls (e.g., not getting paid for work done), and solutions to these pitfalls. This can be done by private sector providers, specially private sector players well beyond the ones who are currently promoting these opportunities (primarily the educational institutions who offer 3–6 month computer training courses, or the training institutions that teach people how to make money online as their business model). For example, for telecom operators, micro-workers can be a high-revenue market segment since Internet data needs of such workers are higher than the average consumer's and possibly more different (need for higher uploading speeds, instead of download only). These firms (or the workers themselves) can facilitate peer networking which help build the soft skills which help to move workers up the value chain, away from commoditized work on platforms.

We conclude that online freelancing and micro-work presents a growth opportunity for Sri Lanka.

Notes

1. Calculated by the authors based on Central Bank data, 2015 (IT Programming Consultancy and Related Activities: 16,409, Total Gross National Income at Market Price: 10,931,932).
2. Definition for Computer literacy: A person (aged 5–69) is considered as a computer literate person if he/she could use computer on his/her own. For example, even if a 5 years old child can play a computer game then he/she is considered as a computer literate person.

References

A.T. Kearney. 2015. *Emerging Markets—Where the Money Is for Australian Companies*. https://www.atkearney.com.au/documents/8942229/8942900/Emerging+Markets_Media+Release_Approved_.pdf/d87adc33-d637-4ddd-bd22-f38862b7326c.

A.T. Kearney. 2016. *A.T. Kearney Releases 2016 Global Services Location Index (GSLI)*. http://www.atkearney.co.uk/news-media/news-releases/-/asset_publisher/00OIL7Jc67KL/content/id/7170816.

Central Bank of Sri Lanka. 2015. *Annual Report 2015*. Colombo: Central Bank of Sri Lanka. http://www.cbsl.gov.lk/pics_n_docs/10_pub/_docs/efr/annual_report/AR2015/English/17_Appendix.pdf.

Coase, Ronald. 1937. "The Nature of the Firm." *Jstor.Org*. http://www.jstor.org/stable/2626876.

DailyFT. 2014. "ICT Workforce Rises by 50% Since 2010, National ICT Workforce Survey 2013 Reveals." http://www.ft.lk/article/287932/ICT-workforce-rises-by-50–since-2010–National-ICT-Workforce-Survey-2013-reveals.

Department of Census and Statistics. 2012. *Census of Population and Housing 2012*. Colombo: Central bank of Sri Lanka. http://www.statistics.gov.lk/PopHouSat/CPH2011/Pages/Activities/Reports/CPH_2012_5Per_Rpt.pdf.

Department of Census and Statistics. 2015a. *Statistical Pocket Book 2015*. Colombo: Department of Census and Statistics. http://www.statistics.gov.lk/pocket%20book/chap13.pdf.

Department of Census and Statistics. 2015b. *Computer Literacy Statistics—2015 (First Six Months)*. Colombo: Department of census and statistics. http://www.statistics.gov.lk/samplesurvey/ComputerLiteracy-2015Q1-Q2-final%20.pdf.

Gamage, Sujata, and Tilan Wijesooriya. 2012. *"Mapping the Higher Education Landscape in Sri Lanka."* Presentation, LIRNEasia.

Hoßfeld, Tobias, Matthias Hirth, and Phuoc Tran-Gia. 2011. "Modeling of Crowdsourcing Platforms and Granularity of Work Organization in Future Internet." In International Teletraffic Congress (ITC). IEEE. http://ieeexplore.ieee.org/document/6038475/.

Kuek, Siou Chew, Paradi-Guilford, Cecilia Maria, Toks Fayomi, Saori Imaizumi, and Panos Ipeirotis. 2015. *The Global Opportunity in Online Outsourcing*. Washington, DC: World Bank Group. http://documents.worldbank.org/curated/en/138371468000900555/pdf/ACS14228-ESW-white-cover-P149016-Box391478B-PUBLIC-World-Bank-Global-OO-Study-WB-Rpt-FinalS.pdf.

Sri Lanka and Australia Chamber of Commerce (SLACC). 2016. "What Makes Sri Lanka the No.1 Destination for IT-BPM Services?" Presentation.

Sunday Observer. 2016. "IT-BPM Industry Poised for Growth—SLASSCOM." http://archives.sundayobserver.lk/2016/05/08/fin02.asp.

Tholons. 2016. *Tholons 2016 Top 100 Outsourcing Destinations*. Tholons. http://www.tholons.com/TholonsTop100/pdf/Tholons_Top_100_2016_Executive_Summary_and_Rankings.pdf.

United Nations Development Program. 2015. *Human Development Report 2015—Work for Human Development.* Washington, DC: United Nations Development Programme. http://www.hdr.undp.org/sites/default/files/2015_human_development_report.pdf.

World Economic Forum. 2015. "Global Information Technology Report 2015." *World Economic Forum.* http://reports.weforum.org/global-information-technology-report-2015/economies/#indexId=NRI&economy=LKA.

15

Do Municipal Broadband Networks Stimulate or Crowd Out Private Investment? An Empirical Analysis of Employment Effects

Hal J. Singer

15.1 Introduction

In March 2015, the FCC granted the petition of the City of Chattanooga Tennessee to preempt a state law that restricts municipally owned broadband ("muni-broadband") deployment (FCC 2015). As was the case for the *Open Internet Order*, the FCC's *Muni-Broadband Order* was preceded by a direct request from President Obama (White House 2015). Much of the debate concerning this action turned on whether the FCC has the legal authority to preempt state laws that restrict or prohibit muni-broadband development. Some legal scholars argue that the only preemption authority at the FCC's disposal, which derives from section 253 of the 1996 Telecommunications Act, concerns preempting state laws that deter entry for private-sector network deployment (Spiwak 2015). As the Supreme Court ruled in *Nixon v. Missouri*, the issue of preemption "does not turn on the merits of municipal telecommunications services" (Nixon v. Missouri 2004).

The author is a principal at Economists Inc., a senior fellow at George Washington's Institute for Public Policy, and an adjunct professor at Georgetown's McDonough School of Business. The author would like to thank Anna Koyfman for her assistance analyzing the NTIA deployment data. The author has served as a consultant to several Internet service providers in the United States and abroad.

H. J. Singer (✉)
Economists Inc., Washington, DC, USA

H. J. Singer
McDonough School of Business, Washington, DC, USA

© The Author(s) 2018
L. Pupillo et al. (eds.), *Digitized Labor*, https://doi.org/10.1007/978-3-319-78420-5_15

To an economist, however, the merits of the policy should dictate the FCC's decision-making. The agency's legal authority to intervene is essential, but not something that lends itself to economic analysis. In response to the D.C. Circuit's ruling in *Verizon*, which provided a potentially alternative source of preemption authority in section 706, Chairman Wheeler stated that "I believe the FCC has the power—and I intend to exercise that power—to preempt state laws that ban competition from community broadband" (Wheeler 2014).

Setting aside the issue of the FCC's legal authority, an economist can ask whether it makes sense for the FCC to preempt state laws that discourage or prevent entry for muni-broadband projects in the first place. Could a state have any reasonable economic basis for discouraging its municipalities from entering the broadband business? If so, then FCC preemption would tend to undercut those reasonable bases. Moreover, if cost-benefit analysis dictates that the best policy is for the FCC to stay out of these affairs—namely, the cost of municipal intervention (for example, deterring private-sector deployment or diverting funding from other priorities) exceeds the benefits (for example, stimulating local economic activity)—the question of legal authority vanishes.

This chapter examines one purported benefit of government ownership of broadband access facilities—namely, stimulation of local economic activity. In Part I, I briefly review the highlights from the economics literature, beginning with what economists have uncovered between private-sector broadband investment and job multipliers (both total multiplier and spillover effects). In contrast, the connection between muni broadband and private-sector employment is harder to find. Economists posit that muni broadband might discourage investment by privately owned ISPs, thereby offsetting any incremental investment from local businesses that exploit the muni network. In Part II, I offer original empiricism that informs this "crowding-out hypothesis." Part III explores the policy implications of these findings.

15.2 Literature Review

The employment effects of capital expenditures in the broadband industry extend beyond the direct employees of the Internet service provider (ISP). "Direct effects" are jobs generated from activities such as installing fiber, while "indirect effects" are job gains associated with communication equipment suppliers. "Induced effects" are the jobs created when the employees of an input provider use their additional income to purchase more goods and services in the local economy. These three effects (direct, indirect, and induced)—collectively referred to as the "total multiplier"—are considered to be the key elements of a traditional analysis of economic impact.

Katz and Callorda (2014) studied the effects of repealing a sales tax exemption in Minnesota on the telecommunications industry. Based on an input-output analysis, they estimate that a $154 million reduction in broadband investment would eliminate 3323 jobs in the state, implying a total job multiplier of 21.6 jobs per million dollars of broadband investment (Katz and Callorda 2014). Audenrode and Sosa (2011) estimated that the effects of reassigning 300 MHz of additional spectrum to mobile broadband would trigger $15.1 billion in new capital spending per year (although the study pertains to mobile broadband, the authors rely on job multipliers derived from wireline services). The authors apply BEA Type II RIMS multipliers to calculate a weighted average of Construction (56%) and Broadcast and Communications Equipment (44%), implying 20.4 jobs for every $1 million invested (Audenrode and Sosa 2011). Using multipliers for telephone apparatus manufacturing (11.8), broadcast and wireless communications equipment (13.8), fiber-optic cable manufacturing (14.4), and construction (26.7), Eisenach et al. (2009) estimated separate multipliers for different types of broadband spending by applying weights to each of the industry multipliers based on the allocation of broadband capital spending to each industry (Singer et al. 2011). Table 15.1 summarizes the relevant literature on the total multiplier effects from broadband investment.

Table 15.1 Summary of total multipliers from broadband investment

Study	Annual investment ($B)	Projected total jobs (000s)	Total multiplier	Method
Crandall and Singer (2010)	30.4	509.5	**16.8**	Multiplier
Audenrode and Sosa (2011)	15.1	307.6	**20.4**	Multiplier
Katz and Callorda (2014)	0.2	3.3	**21.6**	Input-output
Singer and West (2010)	12.7	250.4	**19.7**	Multiplier

Notes Total multiplier is the sum of direct, indirect, and induced effects

Based on the consistency of these estimates, approximately 20 jobs per million dollars of broadband investment is a reasonable predictor of the short-term job impact from a hypothetical broadband deployment.

The total multiplier-based jobs estimate does not account for additional spending in related downstream industries except for those industries that directly benefit from increased spending by broadband input providers. Broadband investment and higher broadband penetration have been shown to create additional, or "spillover" effects in myriad downstream industries, including in healthcare (Meyer et al. 2002) education (Working Party 2009), and energy (Horner 2017), whose ability to enrich and enhance their service offerings is increased by greater availability of broadband internet access (Mandel and Scherer 2012). Broadband spillover effects tend to concentrate in service industries such as financial services and healthcare, yet some have identified an effect in manufacturing as well (Litan et al. 2007). These spillovers have been measured to be roughly equal in magnitude to the direct employment effects generated by broadband investment Katz and Suter (2009). Table 15.2 summarizes the relevant economic literature on spillover effects.

Again, based on the consistency of these findings, it is reasonable to expect a spillover multiplier of slightly over one additional network-induced job per every job created via the total multiplier.

So do muni networks generate the expected employment effects? Muni-broadband deployment has been shown to have no discernible impact on private-sector employment (Deignan 2014). Using a difference-in-difference regression on panel data consisting of 23 years of observations from core-based statistical areas (CBSA), he finds that the private-sector employment effect from muni networks is not statistically significant. To address this paradox, he posits that "physical capital is an important input into the production process, but it does not create economic growth by itself.

Table 15.2 Summary of spillover effects from broadband investment

Study	Annual investment ($B)	Projected total jobs (000s)	Spillover jobs (000s) (spillover multiplier)
Crandall and Singer (2010)	30.4	961.0	452 (**0.89**)
PCIA (2013)	35.5	303.7	194.9 (**1.79**)
Katz and Suter (2009)	6.4	263.9	136.1 (**1.06**)
Atkinson and Schultz (2009)	5.2	498.0	268.5 (**1.17**)

Therefore, public investment plans that focus on end-states, such as attracting a certain business or building a fiber network, are focusing on the inputs of economic growth rather than a root cause, which could end up misallocating resources and encouraging rent-seeking" (Deignan 2014).

Public investment in a service that is competitively provided could perversely discourage future private investment by ISPs, which could have a depressing effect on private employment (Ford 2016). The reason is that publicly owned firms are not profit-maximizers, and thus can be expected to engage in predation (Sappington et al. 2000). From the perspective of an incumbent ISP (or potential entrant), the prospect of competing against a publicly-owned ISP could be sufficient to discourage the next round of investment. Ford notes that "[t]his deterrence effect is particularly pernicious at a time when private providers are undergoing widespread and costly upgrades to their networks. Paradoxically, the resulting lack of private supply may then be used to justify the municipal entry that caused the perceived lack of competition in the first place" (Ford 2016). Accordingly, there can be legitimate economics bases for a state to limit how one city may seek to induce economic migration from another city. As Ford notes, "While it is easy to see a city's leadership wanting to advantage its city over others, it is not clear why the federal and state governments should be complicit in the act" (Ford 2016). Although it might be welfare-reducing on net in cities currently served by private ISPs, muni-broadband may still have a role to play in broadband deployment in markets where private entry is not profitable. Ford concludes that muni-broadband "may be a symptom of the lack of a coherent, economically-informed federal (and state) policy for broadband deployment and adoption in economically-marginal communities" (Ford 2016).

In the FCC's *2015 Preemption Order*, the FCC claimed, without citation to any evidence, that "threat of entry or actual entry of a municipal provider spurs positive responses by the incumbent broadband provider [which] serves the goals of section 706" (FCC 2015). While it is documented that incumbent ISPs react positively (by increasing speeds) to new entry by Google Fiber and other *private* competitors that take profits into consideration when setting prices (Snyder 2015), there is no evidence in the FCC's record to suggest the same reaction will follow muni-broadband investment. Indeed, the FCC acknowledged in its National Broadband Plan that "[m]unicipally financed service may discourage investment by private companies" (FCC 2010). I refer to this theory as the "crowding-out hypothesis."

As noted by Ford, the root cause of any underinvestment in broadband infrastructure is the existence of a positive externality (not captured by ISPs or broadband consumers). ISPs will not deploy to neighborhoods where the private return does not exceed the cost of capital, even when the social return does. More competition in the form of muni-broadband does not treat the problem of underinvestment. To increase the private return, the solution should involve a subsidy to any willing provider, an issue to which I return in Part III.

15.3 Analysis of NTIA Data

To inform the crowding-out hypothesis, I analyzed the Commerce Department's National Telecommunications and Information (NTIA) State Broadband Initiative, which captures deployment data from December 2010 by provider and by download speed. Each observation in NTIA deployment data is at the block-id level, a 15-character FIPS code. The data were aggregated up from the block-id to the county-level based on fastest advertised speed reported by an ISP, which allows one to measure the extent of deployment at a given download speed by a given provider within the county. To focus on the impact of high-speed connections on employment, I omitted observations where the advertised download speeds were less than 10 megabits per second (Mbps).

Next, I categorized a provider-county pair as muni- or privately owned networks based on the name of the provider. For this exercise, I treated any ISP as muni-owned if the provider name in the NTIA data contained any of the following names: "city," "EPB Chat," "Intergovernmental," "North Dakota," or "Tullahoma." The list is potentially under-inclusive, but to the extent that some muni networks are inappropriately categorized as a privately owned network, any differences in the employment effects should be harder to detect. The resulting database yielded 43 provider-county pairs served by a muni-owed ISP and 5817 counties served by a privately owned ISP, respectively, with download speeds in excess of 10 Mbps. Table 15.3 shows the results, broken down by network type and by download speed.

Table 15.3 Number of provider-county pairs by ownership type, by download speed (December 2010)

Download speeds	Muni networks	Private networks
Greater than 10 Mbps and less than 25 Mbps (7)	9	3896
Greater than 25 Mbps and less than 50 Mbps (8)	2	525
Greater than 50 Mbps and less than 100 Mbps (9)	4	806
Greater than 100 Mbps and less than 1 Gbps (10)	18	233
Greater than 1 Gbps (11)	10	357

Notes Numbers is parenthesis correspond to NTIA speed classifications

As of December 2010, there were only ten counties that were served by muni networks capable of reaching 1 gigabit per second (Gbps) download speeds; by comparison, 357 counties enjoyed such speeds from privately owned ISPs. Nearly 3900 counties were served by a privately owned ISPs with speeds between 10 and 25 Mbps.

Finally, I merged this county-level 2010 deployment data with county-level nonfarm private-employment data in 2010 and 2013 from the Bureau of Economic Analysis (BEA 2013). For each county, I computed the cumulative average growth rate (CAGR) in private-sector employment from 2010 to 2013. Table 15.4 shows the results for the sample of privately owned networks.

For these private-sector deployments, one cannot detect any economically significant divergence from the sample average CAGR in private employment growth (1.17%) until the network reaches download speeds of 50 Mbps (9–10 bin); in those counties, private-employment grew at 1.34%. The Gigabit counties (11+bin) appear to enjoy faster job growth relative to the average (1.28%), but not noticeably different from counties served by 50 Mbps. The observed positive correlation between privately owned network speeds and private-sector employment growth does not imply that deployment causes job creation. Causation could be going in the other direction—that is, ISPs may be choosing to deploy a fast network in a city based on an informed guess of future employment growth. Teasing out the direction of causality is beyond the scope of this chapter, but could involve a two-stage regression model in which an instrument is used to predict broadband deployment in stage one.

Table 15.5 replicates the above analysis for counties served by muni networks as of December 2010. Because there were only 34 such counties with access to speeds above 25 Mbps, I combine speed categories before computing the average CAGR; else the average CAGR for a given category would be based on too few observations. Unlike the results for private-sector deployment in Table 15.4, one cannot detect any economically significant

Table 15.4 2010–2013 private-sector employment growth for counties served by a privately owned network provider in 2010

Low speed	High speed	CAGR (%)
7	8	1.12
8	9	1.25
9	10	1.34
10	11	1.27
11	+	1.28
ALL		1.17

Table 15.5 2010–2013 private-sector employment growth for counties served by a muni-owned network provider in 2010

Low speed	High speed	CAGR (%)
8	11	0.96
9	11	0.96
10	11	1.00
ALL		1.17

lift, relative to the sample average of 1.17%, in trailing nonfarm-private employment in these counties. Indeed, trailing private-sector employment growth in muni-served counties appears to underperform the sample average CAGR. Once again, discerning causation is difficult. Hard-hit municipalities (say, with high unemployment) could be selecting public deployments to spur job creation, which would create the false impression that muni networks undermine job growth in a simple correlation analysis. But by using the same metric to gauge employment effects, Tables 15.4 and 15.5 combined lend support to the crowding-out hypothesis. Clearly, more econometric inquiry is needed.

15.4 Policy Implications

The sizable total-employment and spillover effects estimated in the literature suggest that the broadband industry should be nurtured and encouraged. Rather than subjecting ISPs to "light-touch" common-carrier regulation, which raises the specter of rate regulation and unbundling and thereby potentially discourages investment, regulation should be truly light-touch; episodes of discriminatory conduct should be assessed on a case-by-case basis, while bans on new business models and non-discriminatory pricing strategies should be avoided.

Relative to the socially optimal level of broadband investment, the private sector will likely underinvest in the presence of positive externalities, as implied by the significant spillovers. Accordingly, a subsidy on buildout costs (for example, a tax credit for fiber) or a demand-based subsidy (for example, covering the expense of broadband for low-income households) is in order. Barring ISPs from participating in the value created for edge providers, by setting the price of interconnection and paid priority to zero, perversely exacerbates underinvestment caused by externalities.

With respect to the wisdom of government ownership, because muni networks do not appear to generate the same private-sector employment effects as privately owned networks, municipalities cannot cite private employment gains as a benefit of government provision. The finding here is consistent with prior findings in the literature, and consistent with the crowding-out hypothesis that muni networks, which by construction are not profit-maximizing, discourage privately owned networks.

This is not to say that there are no benefits of muni networks. In the absence of any network, a muni network could stimulate economic development and permit residents to develop valuable skills. And there is evidence that muni networks stimulate public employment; some public employment is better than no public employment. Yet public employment can also be stimulated through roads and bridges. Thus, the relevant policy question is how best to spend public resources. Economics counsels that public resources should be allocated to, among other things, *public goods*, such as national defense or lighthouses, which will be under-provided by private parties due to their non-excludable nature. But because broadband (like satellite television) is excludable via a pricing mechanism, including congestion pricing, broadband is closer to a *club good*, which can be profitably provided albeit at significant markups over marginal cost (to cover the large upfront costs). Again, a subsidy that moves broadband adoption toward the socially optimal level (accounting for the positive externalities) is the best course under these circumstances.

Finally, statewide obstacles to funding muni networks could serve as a way for cities to temper their demand for new networks, in the same way that states would prefer that cities temper their demand for new sports stadiums. Muni-broadband should be a last resort for municipalities that cannot be served profitably by private ISPs.

References

Atkinson, Robert, and Ivy Schultz. 2009. *Broadband in America: Where Is It and Where Is It Going?* Preliminary Report Prepared for the Staff of the FCC's Omnibus Broadband Initiative, November. http://www8.gsb.columbia.edu/rtfiles/citi/91a20123-2501-0000-0080-984f56e8d343.pdf.

Audenrode, Van Marc, and David Sosa. 2011. *Private Sector Investment and Employment Impacts of Reassigning Spectrum to Mobile Broadband in the United States.* Analysis Group [Ebook], August. http://www.analysisgroup.com/uploadedfiles/content/news_and_events/news/sosa_audenrode_spectrumimpactstudy_aug2011.pdf.

BEA. 2013. CA25 N Total Full-Time and Part-Time Employment by NAICS Industry 1, Non-Farm Employment (2010 and 2013).

Crandall, Robert W., and Hal J. Singer. 2010. *The Economic Impact of Broadband Investment*, Prepared for Broadband for America, February. http://internetinnovation.org/files/special-reports/Economic_Impact_of_Broadband_Investment_Broadband_for_America_.pdf.

Deignan, Brian. 2014. "Community Broadband, Community Benefits? An Economic Analysis of Local Government Broadband Initiatives." PhD, George Mason University.

Eisenach, Jeffrey A., Hal J. Singer, and Jeffrey D. West. 2009. *Economic Effects of Tax Incentives for Broadband Infrastructure Deployment* (prepared for Fiber-to-the-Home Council).

FCC. 2010. National Broadband Plan. http://hraunfoss.fcc.gov/edocs_public/attachmatch/DOC-296935A1.pdf.

FCC. 2015. In the Matter of City of Wilson, North Carolina, Petition for Preemption of North Carolina General Statute Sections 160A-340 et seq.; The Electric Power Board of Chattanooga, Tennessee, Petition for Preemption of a Portion of Tennessee Code Annotated Section 7-52-601, FCC 15-25, Memorandum Opinion and Order, 30 FCC Rcd 2408.

Ford, George. 2016. *The Impact of Government-Owned Broadband Networks on Private Investment and Consumer Welfare*. State Government Leadership Foundation [Ebook]. http://sglf.org/wp-content/uploads/sites/2/2016/04/SGLF-Muni-Broadband-Paper.pdf.

Horner, Justin. 2017. *Telework: Saving Gas and Reducing Traffic from the Comfort of Your Home*. Mobility Choice [Ebook]. http://mobilitychoice.org/MCtelecommuting.pdf.

Katz, Raul, and Fernando Callorda. 2014. *Assessment of the Economic Impact of the Repeal of the Tax Exemption on Telecommunication Investment in Minnesota*. Saint Paul, MN: Telecom Advisory Services, LLC [Ebook]. http://www.mncca.com/doc/minnesota-study-final-version.pdf.

Katz, Raul, and Stephan Suter. 2009. *Estimating the Economic Impact of the Broadband Stimulus Plan*. NTIA [Ebook]. http://www.elinoam.com/raulkatz/Dr_Raul_Katz_-_BB_Stimulus_Working_Paper.pdf.

Litan, Robert, Robert Crandall, and William Lehr. 2007. "The Effects of Broadband Deployment on Output and Employment: A Cross-Sectional Analysis of U.S. Data." *Brookings*. http://www.brookings.edu/research/papers/2007/06/labor-crandall.

Mandel, Michael, and Judith Scherer. 2012. *The Geography of the App Economy*. South Mountain Economics LLC [Ebook]. http://files.ctia.org/pdf/The_Geography_of_the_App_Economy.pdf.

Meyer, Marlis, Rita Kobb, and Patricia Ryan. 2002. "Virtually Healthy: Chronic Disease Management in the Home." *Disease Management* 5 (2): 87–94. https://doi.org/10.1089/109350702320229186.

Nixon v. Missouri Municipal League, 541 U.S. 124. 2004.

Perce, Alan, Richard Carlson, and Michael Pagano. 2013. *Wireless Broadband Infrastructure: A Catalyst for GDP and Job Growth 2013–2017*, PCIA, September. http://www.pcia.com/images/IAE_Infrastructure_and_Economy_Fall_2013. PDF.

Sappington, David E.M., J. Gregory Sidak, and John R. Lott. 2000. "Are Public Enterprises the Only Credible Predators?" *The University of Chicago Law Review* 67 (1): 271. https://doi.org/10.2307/1600331.

Singer, Hal, and Jeffrey West. 2010. *Economic Effects of Broadband Infrastructure, Deployment and Infrastructure for Broadband Deployment* (prepared for Fiber-to-the-Home Council). http://neoconnect.us/wp-content/uploads/2015/09/Economic_Effects_of_FTTH.pdf.

Singer, Hal, Jeffrey West, and Jeffrey Eisenach. 2011. *Economic Effects of Tax Incentives for Broadband Infrastructure Deployment*. Fiber-to-the-Home Council [Ebook]. https://www.fiberbroadband.org/p/cm/ld/fid=44&tid=76&sid=67.

Snyder, Bill. 2015. "Google Competition Makes AT&T Cut Cost of Gigabit Service in Some Areas." *CIO*. https://www.cio.com/article/2988881/consumer-electronics/google-competition-makes-att-cut-cost-of-gigabit-service-in-some-areas.html.

Spiwak, Lawrence. 2015. "Why the FCC Can't Preempt States on Muni-Broadband." *Bloomberg DNA*. http://www.phoenix-center.org/oped/Bloomberg BNAMuniBroadbandPartII20February2015.pdf.

The White House. 2015. "Broadband That Works: Promoting Competition & Local Choice in Next-Generation Connectivity." https://obamawhitehouse. archives.gov/the-press-office/2015/01/13/fact-sheet-broadband-works-promoting-competition-local-choice-next-gener.

Wheeler, Thomas. 2014. Remarks Before the National Cable & Telecommunications Association. http://transition.fcc.gov/Daily_Releases/Daily_Business/2014/db0430/DOC-26852A1.pdf.

Working Party on Communication Infrastructures and Services Policy. 2009. Network Developments in Support of Innovation and User Needs, Organization for Economic Cooperation and Development. http://www.olis.oecd.org/olis/2009doc.nsf/LinkTo/NT0000889E/$FILE/JT03275973.PDF.

Index